Utilisation and conservation of farm animal genetic resources

Utilisation and conservation of farm animal genetic resources

edited by:
Kor Oldenbroek

Photos cover:
Background: Nordic Gene Bank Farm Animals
Groningen Horses: Hinke Fiona Cnossen, SZH
Scottish Blackface Sheep: Kor Oldenbroek
Finnish Landrace Cockerel: Tapio Tuomela, MTT
Dutch Landrace Goats: Kor Oldenbroek

ISBN: 978-90-8686-032-6

First published, 2007

Wageningen Academic Publishers
The Netherlands, 2007

Preface

The genetic diversity comprised in farm animal species and breeds is an important resource in livestock systems contributing to food supply. For several reasons, within the different species used for food production, only a few breeds are developed towards high-output breeds fitting in high-input systems. In this process many breeds are set aside from the food producing livestock systems. These breeds will be faced with extinction unless new functions for these breeds are found. This is a real threat for the genetic diversity within species. In farm animal species a substantial amount of genetic diversity exists within the breeds. In commercial populations this diversity may be threatened by applying high selection intensities.

In 1992 the second United Nations Conference on Environment and Development in Rio de Janeiro recognised the importance of the diversity in farm animal genetic resources in Agenda 21 and in the Convention on Biological Diversity. This convention raised global awareness for the conservation and sustainable utilisation of genetic resources, also in the group of stakeholders involved in farm animal genetic resources: farmers and their organisations, governmental organisations, breeding companies, education and research institutions and organisations of hobby breeders. This leads to increased interest and activities in the sustainable utilisation and conservation of threatened breeds of livestock.

This book is intended to give insight into the issues of the utilisation and conservation of farm animal genetic resources towards a broad group of readers interested in these subjects. The insight is presented as applications of population, molecular and quantitative genetics that can be used to take appropriate decisions in utilisation and conservation programmes. The book might also be used as teaching material in student courses. Some chapters can be used as an introduction to the issues in BSc courses, while the more technical chapters better fit in MSc and PhD courses. The first two chapters indicate the decisions to be made in utilisation and conservation, chapter 3 surveys the different ways in which the diversity we observe within a species can be characterised, chapter 4 illustrates recent results using this theory for utilisation and conservation purposes, the chapters 5, 6 and 7 give theoretical backgrounds necessary to make decisions and the chapters 8 and 9 present the operation and practical implications of selection and conservation schemes.

In 1998 a group of authors wrote the book "Genebanks and the conservation of farm animal genetic resources" edited by J.K. Oldenbroek to conclude an EU project about the role of conserved genetic material in European animal breeding programmes. A slightly different group decided in 2006 to rewrite that book, because substantial progress was made in the period between 1998 and 2006 in the development of methods and concepts to be used not only in the conservation of farm animal genetic resources, but also in the sustainable utilisation.

As a result this book aims to present applications at the global level instead of being restricted to applications in the EU. All chapters are updated and new information is integrated. In addition to the book of 1998 this book includes a special chapter on the concept of genetic diversity, a review of recent literature on domestication, breed distances and utilisation, a chapter with the theory and applications of genetic contributions and finishes with a chapter on practical implications of utilisation and sustainable management of farm animal genetic resources.

Acknowledgements

The writing of this book is an initiative of the Centre for Genetic Resources, the Netherlands and it is financially supported by the Dutch Ministry of Agriculture, Nature and Food Quality.

Table of contents

Chapter 1. Introduction

Kor Oldenbroek
Centre for Genetic Resources, the Netherlands, P.O. Box 16, 6700 AA Wageningen, The Netherlands

Questions that will be answered in this chapter:

- *What are the challenges for food production in livestock systems?*
- *What changes in livestock systems are underway?*
- *What are the consequences for the use of the different species and breeds?*
- *Why is farm animal genetic diversity important?*
- *Which initiatives are taken at global and regional levels to stop genetic erosion?*
- *What are opportunities and threats for farm animal genetic resources?*
- *Which conservation methods can be applied?*
- *Who are the stakeholders and what activities do they develop?*

Summary

This chapter describes the challenges for food production by livestock brought about by the growth of the global human population and changes in its consumption profile. These challenges result in an intensification of livestock systems, in the development by breeding of a limited number of breeds within a species and in high selection intensities within these breeds. The genetic variation, comprised of components between and within breeds, is under threat. The threat of genetic erosion and the causes are different between species. Genetic diversity should be conserved to maintain the flexibility of livestock systems and to sustain the further development of rural areas. The global and regional activities for conservation and sustainable utilisation of farm animal genetic resources are briefly summarised and methods are outlined. Different groups of stakeholders can play a role in neutralising the threats and using opportunities to utilise and conserve genetic diversity.

1. Challenges for food production in livestock systems

The major challenge for food production in agriculture is the ongoing growth of the human population from 6 billion people today towards 9 billion in 2050. The first Millennium Goal adopted by the United Nations in 2000 is to eradicate poverty and hunger. Five years after the adoption of this goal it can be concluded that there has been some successes, led by Asia by developments in China and India (United Nations, 2005). But millions of people have sunk deeper into poverty in Sub Saharan Africa and in South Asia, where half of the children under the age of 5 are malnourished.

Livestock systems play an important role in agriculture by producing high quality food. In the developed countries the higher welfare status is accompanied with a higher consumption of meat, milk and eggs. In developing countries some 70 percent of the world's rural poor (2 billion people) depend on livestock as an important component of their livelihoods (Hoffmann and Scherf, 2005). There, animals provide not only meat, milk and eggs, but also fibre, fertiliser for crops, manure for fuel and draught power. Moreover, they fulfil social and cultural functions in these livelihoods such as dowry, savings, gifts and ceremonies.

The challenges for food production in livestock systems in the developed countries are food quality and food safety to safeguard human health, animal welfare in intensive systems and sustainable use of resources. The challenges in developing countries are focused on the increase of production of milk, meat and eggs to close the big gap between demand and production. The utilisation and management of farm animal genetic resources contribute to meet these different challenges in developed and developing countries.

2. Changes in livestock systems

Between 1995 and 2004 global milk production increased by 15 percent, egg production by 35 percent and meat production by 25 percent (Rosati *et al.*, 2005). The growth in production is predominantly realised in countries with a rapidly growing livestock sector: Brazil, China, Mexico, Thailand and several East European countries (The World Bank, 2005). In general these increases in production are realised by intensification of livestock systems towards high-input high-output systems. The genetic resources for these intensive production systems are only a few breeds and lines, which are developed by a limited number of multinational breeding companies. Continuously, many breeds and recently developed breeds and lines are set aside from the primary food production chains in the intensification of livestock production and in the global concentration of breeding activities. To get an impression of the threats for farm animal genetic resources, FAO prepared a "Report on the State of the World's Animal Genetic Resources". Individual country reports, describing the livestock systems and the use and conservation of breeds, were written by governments at a request of FAO willing to describe the state of farm animal genetic resources in order to facilitate national, regional and global action plans for sustainable utilisation and conservation. Up to 2006, 169 country reports are submitted (FAO, 2006).

In an analysis of 148 of these country reports available mid 2005 (Oldenbroek, 2006) large differences between continents in developments of livestock systems were found and it could be concluded that livestock systems are very dynamic at the moment all over the world. Four global regions were used in the analysis: Africa, Asia, Europe and the "New World". The latter region comprises the Americas and the South West Pacific.

The strong population growth of the African and Asian human population requires in these continents a 200% increase in food production within 15 years (Hoffmann and Scherf, 2005). In many of these countries a strong pressure on land exists. This all requires intensification of food producing systems and, especially, a strong improvement of the quality of the animals used in livestock systems. In general, in Asia this process of genetic improvement and intensification is faster than in Africa. Many African countries are fighting against chronic poverty and the high incidence of infectious human diseases, including AIDS, hamper economic development. In Europe the introduction of environmental and production restrictions increased production costs, decreased the self sufficiency and induced changes in livestock systems. A substantial amount of land is no longer used for agriculture and is giving back to nature. Less intensive systems like organic farming are introduced and growing in importance. At the same time a significant number of part-time farmers and hobbyists keeps farm animals in rural areas. In the New World the farming structure changed significantly towards large farms with many animals driven by economic growth, a high efficiency of production and requirements of export markets. These forces stimulate the intensification of animal production and require a high protection of the health status in all parts of the production chains. There industrialised animal production dominates, but in the less developed countries in the New World subsistence farming is still important.

3. Consequences for the use of species and breeds

The perspective for a breed depends to a great extent of its present and future function(s) in livestock systems (Oldenbroek, 2006). A change in livestock systems may have great impact on the use of breeds. Livestock system development is driven by many external and internal factors like:

- the presence of ecosystems suitable for animal production;
- the country's policies for the use of animals;
- the prevalence or outbreak of diseases;
- the political (in)stability;
- the available infrastructure;
- the possibilities for introduction of exotic breeds;
- the growth of the human population;
- the growth of the country's economy;

- the training of human resources;
- the possibilities to invest money in livestock systems and breed improvement;
- the market and export possibilities for livestock products.

Therefore, in the strategies for utilisation and conservation of farm animal genetic resources a lot of attention has to be paid to the role of genetic diversity in livestock systems and the changes in the livestock systems to be expected. The position and the expected changes indicate which genetic resources will be used in the (next) future and which has to be conserved.

At the global level large differences in the present and future use within livestock systems and in the conservation of the six important farm animal species (cattle, sheep, goats, pigs, chicken and horses) are found (Oldenbroek, 2006).

3.1. Cattle

For high-input systems, specialised breeds of dairy or beef cattle are developed through intense selection and their genetic material is widely disseminated. Nucleus breeding has started in dairy cattle, but there remain many dairy farmers who participate in breeding activities. On the global level, an intense selection for a few production traits and a large exchange of semen from the best bulls has led to low effective population sizes in the most popular dairy breeds, with a real risk of losing genetic diversity within the breeds involved.

In the dairy sector, the Holstein Friesian breed dominates and in the beef sector French beef breeds are likely to obtain a similar position in the future and take over the position of the British breeds. In many countries, these specialised breeds are used for cross-breeding to improve the performance of local breeds. Only in a few situations stable crossbreeding systems have been developed, in which populations of the local breeds are not only used but also conserved. Nowadays, in some countries, multi purpose cattle breeds are used for organic farming, for new functions like landscape and nature management, or are kept as suckler cows (cows which raise calves) by hobbyists. All over the world conservation programmes have to be developed for local cattle breeds and for multi-purpose breeds that will no longer be used for their original functions. Artificial reproduction techniques (including cloning) and cryoconservation techniques are very well developed in this species, facilitating conservation.

3.2. Sheep

In countries with high input livestock systems in Europe, North America and Australia the number of sheep has declined in recent years. Sheep wool now has a lower economic value, and this is a threat to some breeds. In North and West Europe the future use of this species seems to be nature management. This offers a great opportunity for *in vivo* conservation programmes, because large flocks are required for this purpose, which is attractive for conservation. In small scale farming systems and the semi-intense systems in Africa, Asia and East and South Europe sheep are still important for meat or milk production. In some religions sheep have a ceremonial function. These functions guarantee continued utilisation of the species. In general natural mating systems are applied in sheep breeding. Artificial reproduction techniques are developed, but only in a few countries they are used as a routine for example in sire reference schemes to improve breeding value estimates.

3.3. Goats

The goat's importance in small scale farming systems for milk and meat production, its stable numbers and the wide variety of conditions under which goats can be kept are guarantees of continued utilisation. This species is not faced with real threats. In general natural mating systems are applied in goat breeding. Artificial reproduction techniques are developed, but only applied experimentally.

3.4. Pigs

In Europe, North America and Australia pork production is dominated by a few multinational companies. In the concentration of the breeding industry, continuously many breeds and lines become uneconomic for market driven food production. In a number of regions including Europe, Africa and North America, relatively few local pig breeds exist. Conversely, in East Asia many local pig breeds can be found. The speed of industrialisation and specialisation, in combination with the lack of opportunities for the *in vivo* conservation of pigs means that this species requires special attention in national and regional conservation programmes. Frozen semen is used for the dissemination of genetic improvement and frozen semen and frozen embryos are used for exchange of genetic material within the companies between their populations present in different countries.

3.5. Chickens

The breeding and production of this species is the most specialised and industrialised of all animal species. It is showing similarities with plant breeding and production. At a global level only a very few multinationals are active selling highly specialised hybrid layers and broilers. The number of chickens increases very fast at the global level, mainly due to active marketing by the layer and poultry industry. In developing countries the substantial role of chickens in small-scale farming, and the preference of local people for meat from local chicken breeds will stimulate the use of these local breeds in the future. In the developed world many people keep chickens as a hobby, which maybe an opportunity for conservation. Artificial reproduction techniques are well developed, but cryoconservation of embryos or primordial germ cells still has to be developed.

3.6. Horses

In the past, horses were mainly used for draught and transport. The onset of the mechanisation in transportation and later on in agriculture has meant that in many parts of the world horses are bred almost entirely for hobby and sport purposes. It develops itself as an industry in which hobbyists play an important role. The large variety in purposes may stimulate the maintenance of a wide genetic diversity within the species. However, in general, the genetic diversity within horse breeds is threatened by the wide use of a few popular stallions. The "heavy" breeds, originally bred for draught purposes, are threatened. In some countries they are still kept for meat production. Artificial reproduction techniques (including cloning) and cryoconservation techniques are very well developed in this species, facilitating conservation.

3.7. The threat of genetic erosion

From the former subsections it can be concluded that the threat of genetic erosion differs between the species. At the global level the threats are of little importance in goats and most severe in pigs. In pigs, chicken and cattle only a few breeds or lines are developed towards high-output breeds fitting in high-input systems. In this process many breeds are set aside from the food producing livestock systems. These breeds will be faced with extinction unless new functions for these breeds are found. This is a real threat for the genetic diversity within species. In farm animal species a substantial amount of genetic diversity exists within the breeds (chapter 3). In commercial populations of cattle, pigs and chicken this diversity may be threatened by applying high selection intensities.

4. The importance of farm animal genetic diversity and objectives for conservation

The genetic diversity within a farm animal species is the resource to realise required changes in the phenotypic characteristics of a population. These characteristics can roughly be categorised in production traits (quantity and quality) and fitness traits (adaptation, conformation, fertility and disease resistance). Genetic variation in production traits is the base for artificial selection applied in breeding programmes to realise genetic improvement in forthcoming populations. Genetic variation in fitness traits may be affected by artificial selection in breeding programs: for example selection for milk production has a negative effect on fertility of dairy cows. The genetic variation is the base for natural selection that facilitates the adaptation of a population to its (changing) environment and the base for artificial selection in commercial populations. Besides being sources of genetic variation, breeds can bear other values. Following these considerations, objectives for conservation are distinguished in the following subsections. The first four aim to maintain opportunities for the future, while the last three aim at present and future utilisation.

4.1. Opportunities to meet future market demands

In the prosperous countries of the World the demand for specialised food from animal origin increases. This results in a diversification of animal production systems and of animal products. Besides, prosperity increases the use of animals for other goals like hobby farming and the use of animals for sports (horses). These developments request a large variability in the genetic variation of the species used. The few breeds currently used worldwide in the high-input high-output production systems can not meet all these future market demands.

4.2. Insurance against future changes in production circumstances

The high-input high-output systems are characterised by the use of high levels of fertilisers and concentrates. Within these systems veterinary treatment with drugs for preventive and clinical use are sometimes practised at high levels. Agricultural pollution and resistance against drugs can create conditions for animal production in which higher levels of feed intake or disease resistance are required. New diseases could also arise following the increasing exchanges of biological material among areas of the world and possible climate changes. Conservation of genetic variation is necessary as an insurance against changes in production circumstances or the threats of new diseases.

4.3. Insurance against the loss of resources with a high strategic value

Recent outbreaks of diseases, wars and the development of biological weapons created political awareness for the value of genetic resources currently in use for food production. When these resources are destroyed, the investments in it are lost and breeding material should be obtained from abroad. A back up of the genetic material currently in use, preserved in a gene bank, gives the opportunity for a quick restart of a breeding program and safeguards national food production systems.

4.4. Opportunities for research

World-wide in animal production, molecular geneticists are searching for genes, which influence production, quality of products, and health and reproduction traits of animals. In this search, the analysis of a high variety of breeds and crosses between breeds with extreme characteristics has played an important role. The latter guarantees a high degree of heterozygosity and linkage disequilibrium, which is required to detect associations between highly polymorphic marker loci and polymorphism's at quantitative and qualitative trait loci.

4.5. Present socio-economic value

In many countries and areas of the world local breeds guarantee a livelihood in harsh areas where high-input high-output livestock systems are not feasible. In addition, the local breeds are used by a small group of farmers sometimes for special reasons (e.g. biological farming or grazing of marginal lands) or special purposes (e.g. local products for niche markets). The development of breeding programs for these local breeds is too costly for breeding organisations and the absence of a breeding program is a direct threat for the existence of the breed. However, present socio-economic value, which creates income for large human populations in developing countries and for small farmers in rural areas of developed countries (like the Alps in Europe) and the renewed interest for the development of regional products, justify the establishment of a conservation program.

4.6. Cultural and historic reasons

Many breeds are the result of a long domestication process and a long period of adaptation to local circumstances. They reflect a long history of symbioses between mankind and farm animals and can help to clarify adaptation processes, which can still be worthwhile for the management of animals in present production systems. More generally they are documents of the history of rural populations and as such they can be considered as cultural properties. The history of these breeds can be used in education to illustrate the way of live of mankind in the past.

4.7. Ecological value

Within the developed world the awareness is growing for the ecological value of regions as a result of landscape, nature and farm management. Within this complex the presence of animals from native origin which interact with parts of this complex is of great ecological importance. Besides, these animals can contribute to the development of local products with an ecological image.

5. History of initiatives to stop genetic erosion at global and regional levels

World-wide the discussion on conservation of genetic resources in animal production started much later than in plant production. However, already at the start of artificial insemination of cattle in the fifties of the 20th century, Swedish AI-studs conserved semen from each bull used for breeding. In the sixties, scientific and farmer communities draw attention to the high rate of erosion of animal genetic resources. In Europe, farmers were leaving the rural areas where much breed diversity was present and many local breeds were replaced by a few highly promoted and intensively selected breeds. These intensively selected breeds were also exported to developing countries outside Europe and replaced breeds which were well adapted to circumstances and management systems deviating sharply from those in Europe. In 1972 the first United Nations Conference on the Environment in Stockholm recognised these developments and problems. Ultimately, the first Global Technical Consultation on Genetic Resources was held at FAO-headquarters in Rome in 1980.

In 1985 FAO introduced under the responsibility of the Commission on Genetic Resources for Food and Agriculture an expanded Global Strategy for the Management of Farm Animal Resources. In 1992 FAO launched a special action program for the Global Management of Farm Animal Genetic Resources with a framework to stimulate national participation in the global effort to implement conservation activities. National and regional Focal Points play an important role in stimulating and co-ordinating these actions and to provide technical guidelines for conservation. The Domestic Animal Diversity Information System (DAD-IS) is used to collect information on breeds and conservation activities and it offers the opportunity to retrieve guidelines for conservation activities. In 1998 it was requested that FAO would co-ordinate the development of a country-driven Report on the State of the World's Animal Genetic Resources. In total 169 country reports were written by the national governments in

2002–2005. The analyses were finished in 2006 (FAO, 2006) and the resulting global and regional action plans will be assessed in 2007.

In 1992 the second United Nations Conference on the Environment and Development in Rio de Janeiro recognised the importance of farm animal genetic resources in Agenda 21 and in the Convention on Biological Diversity. The CBD considers farm animal genetic variation as a component of the overall biological diversity. The CBD recognises the sovereignty of each country over his own genetic resources, which implies also the obligation to conserve these resources and use these resources carefully.

The awareness of scientists and their willingness to develop scientific tools to manage animal genetic diversity can be illustrated by the activities of the European Association for Animal Production (EAAP). In 1980 the EAAP established a working group in this field. The main activities today are to develop the European farm animal biodiversity information system (EFABIS) and to integrate the animal genetic science in conservation activities.

In the past decades citizens and farmers interested in the maintenance of native breeds founded national breed conservancy associations. These nongovernmental organisations initiated a variety of activities to conserve *in situ* native breeds with a cultural historic or ecological value and to draw the attention of other stakeholders for this issue. On the regional level, for example SAVE in Europe, as well as on the global level, Rare Breed International (RBI), umbrella organisations exist.

6. Opportunities and threats for farm animal genetic resources

6.1. Threats

Many threats exist for farm animal genetic resources (Hoffmann and Scherf, 2005):

- Social and economic changes, urbanisation and policy factors leading to intensification of production and the ruling out of many local breeds or recently developed multi purpose breeds.
- Global marketing of exotic breeding material accompanied with the substitution of local breeds by exotics.
- Liberalisation of markets for animal products that hampers the development of local production systems with local breeds.
- Loss of traditional livelihoods and cultural diversity is a direct threat for the existence of local breeds.
- Changes in ecosystems may require other adaptive capacities of the animals involved.

- Wars, political instabilities, diseases and natural disasters destroy not only populations of local breeds but often the whole infrastructure for breeding is lost.

6.2. Opportunities

Besides threats, several opportunities can be found to use farm animal genetic resources (Oldenbroek, 2006):

- When a breed or a line is exploited in a viable livestock system it is often managed and developed by a breeding company or an organisation of breeders. In the modern breeding schemes the conservation of the genetic diversity can and should be taken into account and should be optimised in combination with selection for the desired traits. These optimisation techniques are well developed and effective and will be described in chapter 8.
- Grazing animals, particularly local and well adapted breeds of sheep, cattle and horses can play an important role in nature management. Where appropriate, this role offers a great opportunity for the conservation of the herbivore species as large numbers of animals are potentially involved.
- The development of organic farming offers an opportunity for the conservation of the recently developed dual purpose breeds. In many cases these breeds are set aside from the intensive livestock systems. However, they fit better in the production goals of organic farming than intensively selected breeds or crossbreds.
- The development and production of special regional products in natural environments for niche markets offers the possibility to use native breeds and to make them profitable again.
- Hobbyists play a very important role in the utilisation and conservation of the between breed variation in chicken, horse, sheep, goat and cattle.

7. Conservation methods to be applied

Theoretically, three types of conservation can be applied (FAO, 2006):

- *In situ* conservation, defined as conservation of livestock through continued use by livestock keepers in the agro-ecosystem in which the livestock evolved or are now normally found (includes breeding programmes). This method of conservation is to be preferred. All objectives of conservation can be reached the best and it offers possibilities for utilisation. Besides, the development of the breed can continue and it facilitates adaptation to changing circumstances. However, the risks of inbreeding and random drift have to receive full attention in the breeding schemes of these often small populations.
- *Ex situ in vivo* conservation, defined as conservation through maintenance of live populations not kept under normal farm conditions and/or outside of the area in

which they evolved or are now normally found. For cultural historic reasons only a few animals of a breed are kept in zoos or farm parks were they fulfil a museum role. The costs of this type of conservation are low, but the breed is kept outside its environment and further adaptation to this environment is impossible.
- *Ex situ* (cryo) conservation, defined as the storage of gametes of embryos in liquid nitrogen. An overview of the literature (Hiemstra *et al.*, 2006) indicates that for most farm animal species it is possible to cryoconserve semen and realise high or acceptable levels of conception after thawing the semen and inseminating females. For many farm animal species frozen embryos can be used to create live offspring. Also, developments have been made in freezing techniques for oocytes. For all animal species DNA-storage and storage of somatic cells is a well-known technology. However, techniques like nuclear transfer should be developed further and more efficient in order to use these types of storage to regenerate animals after conservation.

In practice, the difference between *in situ* conservation and *ex situ in vivo* conservation can be rather vague and only a clear distinction can be made as: *in vivo* (the combination of *in situ* and *ex situ in vivo)* and *in vitro (ex situ)* conservation. Integration of *in situ* and *ex situ* methods can provide a powerful conservation strategy, as we will see in chapters 2 and 8.

8. Stakeholders for *in vivo* and *in vitro* conservation programs

At the global level many stakeholders are involved in the conservation of farm animal genetic resource: national governments, institutes for research and education (including universities), non governmental organisations, breeders' associations, farmers and pastoralists, part time farmers and hobbyists, and breeding companies. The following section provides a brief overview of the role of the various stakeholders.

8.1. National governments

National governments provide the legal base for utilisation and conservation programmes. This is done under legislation relating to the protection of biodiversity or under legislation regulating the management of farm animal genetic resources, of livestock production and of animal breeding. The national governments should be heavily involved in the development of national strategies for management, utilisation and conservation of farm animal genetic resources and they should provide funding for implementing national strategies (chapter 9).

In some African and Asian countries, national governments are involved in breeding activities, often with the aim of increasing national self sufficiency in food of animal origin. In most cases they own nucleus farms, where local or exotic cattle are kept. These nucleus farms sell breeding stock (males) to improve populations owned by (small) farmers. This system plays an important role in the utilisation and conservation of these breeds. The farmers keep large numbers of production animals and the nucleus farms take care of the genetic diversity of the populations.

In a number of European countries government policies are increasingly focused on conservation and landscape enhancement in rural areas where the economic viability of farming is limited. Ruminants can play a role in these policies. In parts of Europe, governments are also motivated to maintain livestock breeds for socio-economic or cultural/historic reasons.

There are many types of governmental institutions such as therapeutic farms, prisons, demonstration farms, farm parks, and museums at which local breeds may be kept. The number of animals conserved in such locations is generally low, leading to risks of inbreeding and random loss of alleles with a low frequency in the population.

8.2. Education and research institutes

Farms linked to universities and research institutes are often involved in selling breeding animals or conserving local breeds. They combine these activities with their primary tasks of educating students and carrying out research. Many universities and research institutes try to conserve locally developed breeds, which are no longer used by the industry. They pay a lot of attention to the maintenance of the genetic diversity within these populations. Universities and institutes are motivated for these activities as users of genetic diversity in their basic research to unravel genetic and physiological processes with genomic techniques.

8.3. NGOs, part-time farmers and hobbyists

In many developed countries nongovernmental organisations (NGOs) conserve and stimulate the use of local breeds by (part-time) farmers and hobbyists. These NGO's and their members play an important role in keeping local breeds of chicken, horse, sheep, goat and cattle. One of their drives is to demonstrate the cultural and historic aspects of the different breeds for the purpose of education and recreation, or to produce special products for niche markets. In general their knowledge in genetic management is limited and the participation of individual breeders in breeding and conservation programmes is often on a voluntary basis. The number of part-time farmers and hobbyists keeping farm animals is increasing in the Europe and in North America and the Southwest Pacific regions. Most livestock species, except the pig, are kept for hobby purposes.

8.4. Breeder's associations, farmers and pastoralists

In Europe and in North America many breeders' associations exist. Together with the farmers involved they try to take advantage of niche markets to sell speciality products from local breeds, often kept in natural environments. In these circumstances the local breeds are an integral part of the brand, and this provides an opportunity for profitable production using breeds that would otherwise be uneconomic. In many countries farmers or farmer's organisations have become involved in organic farming. In some cases, traditional breeds are favoured in organic systems because of their good adaptation to the management conditions, and for marketing reasons. Potential opportunities to export organic products are increasingly recognised in many East European countries. In a number of African countries the use of local indigenous animal genetic resources within the traditional low external input production systems is considered to be the form of utilisation and conservation which best suits the local conditions and avoids problems related to the lack of financial resources for other forms of conservation. Uncontrolled mating, changes of production systems and cross-breeding are the significant risks in this form of utilisation and conservation.

8.5. Breeding companies

Primary food production in the developed world tends to follow an integrative approach with all participants in the production chain from breeding companies, suppliers of equipment, feed suppliers, veterinarians and processing industry towards retailers and consumers. Their primary focus is on uniformity of the product and the production methods within the chain. In itself, this is a threat for the maintenance of genetic diversity. But to be competitive, differentiation between chains and development of new products is required. The required differentiation and development is a real opportunity for the utilisation and management of genetic diversity by breeding companies.

In the poultry industry, only a very few multinationals are actively selling highly specialised hybrid layers and broilers using a very limited number of intense selected lines as basic breeding stock. The number of these specialised chickens producing eggs or poultry meat is increasing very quickly at the global level, mainly as the result of intensive marketing by the layer and poultry industries.

In Europe, North America and Australia, pork production is highly industrialised and a few multinational breeding companies dominate in the production chains. These companies develop a few lines from a limited number of breeds and these lines are used globally. Frozen semen is used for the dissemination of genetic improvement, and frozen semen and frozen embryos are used for to transfer genetic material on an international scale.

Specialised dairy and beef breeding is also a multinational activity where frozen semen and frozen embryos are used to disseminate the genetic improvement obtained in the countries and herds of origin.

In the genetic improvement of the pure lines, breeding companies use sophisticated methods to control the effective population size and to avoid inbreeding. The companies do not want to limit their future scope for selective breeding. Therefore, the genetic diversity within their breeds and lines is safeguarded.

References

FAO, 2006. The State of the World's Animal Genetic Resources for Food and Agriculture, first draft, Rome, Italy.

Hiemstra, S.J., T. van der Lende and H. Woelders, 2006. The potential of cryopreservation and reproductive technologies for animal genetic resources conservation strategies. In: The Role of Biotechnology in exploring and protecting Agricultural Genetic Resources. John Ruane and Andrea Sonnino (eds.), FAO, Rome, Italy.

Hoffmann, I., and B. Scherf, 2005. Management of farm animal genetic diversity: opportunities and challenges. In: Animal production and animal science worldwide. WAAP book of the year 2005. A. Rosati, A. Tewolde and C. Mosconi (eds.), Wageningen Academic Publishers, pp. 221-246.

Oldenbroek, K., 2006. *In situ* conservation strategies; a quick scan of SoW-AnGR country reports. In: J. Gibson, S. Gamage, O. Hanotte, L. Iñiguez, J.C. Maillard, B. Rischkowsky, D. Semambo and J. Toll (eds.), 2006. Options and Strategies for the Conservation of Farm Animal Genetic Resources: Report of an International Workshop and Presented Papers (7-10 November 2005, Montpellier, France) [CD-ROM]. CGIAR System-wide Genetic Resources Programme (SGRP)/Biodiversity International, Rome, Italy.

Rosati, A., A. Tewolde and C. Mosconi, 2005. Section 4, Statistics. In: Animal production and animal science worldwide. WAAP book of the year 2005. A. Rosati, A. Tewolde and C. Mosconi (eds.), Wageningen Academic Publishers, pp. 231-356.

The World Bank, 2005. Managing the Livestock Revolution. Policy and Technology to Address the Negative Impacts of a Fast-Growing Sector. Report No. 32725-GLB, The International Bank for Reconstruction and Development, Washington DC, USA.

United Nations, 2005. The Millennium Development Goals Report 2005, United Nations, New York.

Chapter 2. Strategies for moving from conservation to utilisation

Gustavo Gandini[1] *and Kor Oldenbroek*[2]
[1]*Department of Veterinary Sciences and Technologies for Food Safety, University of Milan, Via Celoria, 10, 20133 Milan, Italy*
[2]*Centre for Genetic Resources, the Netherlands, P.O. Box 16, 6700AA Wageningen, The Netherlands*

Questions that will be answered in this chapter:

- *What are the objectives for the conservation and utilisation of farm animal genetic resources?*
- *Which techniques are available for conservation?*
- *Which options do we have for a sustainable use of local breeds?*
- *Which material should be stored in cryobanks?*
- *Do conservation strategies differ in costs?*
- *How to choose the most appropriate conservation strategy?*

Summary

This chapter first introduces a general framework extending from cryoconservation to sustainable utilisation, in which *in situ* and *ex situ* techniques differ in their capacity to reach the various conservation objectives. For *in situ* conservation different options for maintaining self-sustaining local breeds are discussed. Some relevant aspects of *ex situ* conservation related to the creation of cryobanks, including selection of donor animals and the type and amount of material to be stored, are analysed as a function of the conservation objectives. In reviewing the scarce literature on costs, the chapter provides a general framework to compare costs of the different strategies for animal genetic resources management. Finally, some criteria to choose the most appropriate conservation strategy for a breed in its breeding environment are proposed.

1. Objectives in conservation and utilisation of farm animal genetic resources

The many-faceted character of farm animal genetic resources reflects a variety of possible objectives in their utilisation and conservation by society, and these can be summarised into two main objectives:

1. Flexibility of the genetic system based on different arguments:
   - insurance against changes in market or environmental conditions;
   - safeguard against emerging diseases, political instability and natural disasters;
   - opportunities for research.

In this view farm animal genetic resources are sources of genetic variation of fundamental importance to ensure future genetic improvement, to satisfy possible future changes in the markets and in the production environment, and to safeguard against disasters that give an acute loss of genetic resources.

2. Sustainable utilisation of rural areas:
   - opportunities for development for rural communities;
   - maintenance of agro-ecosystem diversity;
   - maintenance of rural cultural diversity.

In fact, in many parts of the world breeds particularly adapted to extreme environments are unique sources of income for the rural communities. The link between local breeds and the environment where they were developed sometimes makes them important elements of cultural diversity, as they reflect a history of symbiosis of relatively long periods with mankind, and key components of the agro-ecosystems diversity.

A world wide analysis shows that in different areas of the world stakeholders will assign different rankings to these components. In the poor areas, the element of farm animal genetic resources as income tool for the rural communities is of major importance. The cultural component will play a stronger role for example in the European context as historical witnesses and opportunity for rural tourism, while in African countries it can contribute in maintaining the identity of human communities.

To manage farm animal genetic resources appropriately, and to get commitment from the society for conservation, we need to develop parameters to measure the non-conventional services of the breeds, such as their possible cultural and environmental roles. A methodology to measure the cultural dimension of local breeds is reported in Box 2.1. For conventional goods and services, markets provide the information to estimate their economic value (Roosen *et al.*, 2005). However, the cultural, environmental and insurance roles of the resources generally are not recognised by the market, therefore they need to be valued by using specific techniques, as we will briefly discuss in paragraph 3.1.

Box 2.1. A methodology to assess the cultural value of breeds.

A breed can be considered as a cultural property (Gandini and Villa, 2003) in relation to its role as an 'historical witness', and because it is a point of reference in ancient local traditions, and it is therefore a 'custodian of local traditions'.
The historical value can than be analysed as:

- *Antiquity,* as the period from which the breed has been present in the traditional farming area. The longer this period, the greater the impact it had on the rural society.
- *Agricultural systems* historically linked to the breed, including farming techniques.
- *Role in landscape* formation, or as part of the landscape itself.
- *Role in gastronomy,* in the development of typical products and contribution in recipes.
- *Role in folklore,* directly or through farming methods, including religious traditions.
- *Role in handicrafts,* through practices linked to its farming or by furnishing raw materials.
- *Presence in forms of higher artistic expression,* the extent to which the breed has been perceived as a typical component of the rural dimension, in figurative arts, poetry, etc.

Once the value of the breed as historical witness has been confirmed, the value as custodian of local traditions in rural ares can be analysed as:

- *Role in maintaining landscape,* as percentage of farms that contribute to maintaining features of the traditional farming landscape associated with the breed.
- *Role in maintaining gastronomy,* as current linkages between the breed and typical local products or recipes.
- *Role in maintaining folklore,* as recounting of folklore and re-enacting of religious traditions in the area, linked directly or indirectly to the breed.
- *Role in maintaining handicrafts,* as practise of forms of local handicrafts in the area, linked directly or indirectly of the breed.

In the next paragraphs we will see how different techniques allow us to reach the various objectives, to utilise the various components of animal genetic resources, to maintain opportunities and to develop added value to livestock and rural systems.

2. Techniques

Techniques for conservation of animal genetic resources are generally divided into *in situ*, i.e. the utilisation of breeds within their production systems, and *ex situ*. *Ex situ* techniques are further divided, according to FAO (1998), in cryoconservation of genetic material, which includes haploid cells (semen, oocytes) and diploid cells (*in vivo* and *in vitro* embryos, somatic cells), and *ex situ* live, i.e. the maintenance of live animals of a breed outside its production system (e.g. herds kept in protected areas, experimental and show farms, research stations, zoos, by hobby breeders. Note that sometimes there is not a clear distinct line between *in situ* and *ex situ* live, as might be the case in keeping animals of rare breeds by hobby breeders). A wide consensus exists for *in situ* conservation. The Convention of Biological Diversity (CBD, art. 8) emphasises the importance of *in situ* conservation and advises (art. 9) *ex situ* conservation as an essential activity complementary to *in situ* measures. FAO underlines in its 'Guidelines' (FAO, 1998) the priority for *in situ* conservation. The Common Agriculture Policy (CAP) of the European Community provides for incentives to local endangered breeds in their production systems. *Ex situ*, however, continues to provide powerful and safe tools for conservation of farm animal genetic resource. Therefore, it seems reasonable to make efforts to build a framework, and to search for options for an effective integration of *in situ* and *ex situ* techniques, in which *ex situ* conservation is complementary to *in situ* conservation.

In situ and *ex situ* techniques differ in their capacity to achieve the different conservation objectives listed in the first paragraph (Table 2.1). *In situ* conservation is effective in reaching all objectives other than the objective of safeguarding against emerging diseases, political instability and natural disasters. The *ex situ* live method has no opportunity for rural development since this technique removes the breed from its socio-economic context. Also the cultural and ecological objectives cannot be effectively pursued with this conservation method. A reliance on cryoconservation is therefore an option when socio-economic, cultural and ecological values are missing or are of no concern and is the method of choice to safeguard farm animal genetic resources against disasters.

Table 2.1. Conservation techniques and objectives.

	Technique		
Objective	**cryocon'**	***ex situ* live**	***in situ***
Flexibility of the genetic system, as:			
insurance for changes in production conditions	yes*	yes*	yes*
safeguard against diseases, disasters, etc.	yes	no	no
opportunities for research	yes	yes	yes
Sustainable utilisation of rural areas:			
opportunities for rural development	no	no	yes
maintenance of agro-ecosystem diversity	no	poor	yes
conservation of rural cultural diversity	no	poor	yes

*Some differences among the three techniques, see Table 2.2.

In Table 2.2, techniques are compared on the basis of some important factors associated with the conservation objectives: (1) opportunities for breed evolution and genetic adaptation to a changing environment; (2) opportunities to better characterise the breed; (3) exposure of the breed to random genetic drift and inbreeding. Besides freezing the ontogeny process, cryoconservation 'freezes' also the evolutionary process of the breed and impairs its genetic adaptation and therefore possibly reduces its future production capacity. Nowadays, very little is known on the performances of most local breeds, for production and particularly for fitness traits. There is a need to increase our knowledge of local breeds in order to better utilise them and to define their value (see paragraph 3.1). Populations of small size are exposed to random genetic drift, which can be measured as effective population size and is realised as inbreeding. The rate of genetic drift (see chapters 3, 4, 7, 8) can be controlled by tuning the demographic structure of the population, such as the sex ratio of parents and the variance of their reproductive success, and by managed selection and mating of sires and dams. *Ex situ* live populations are expected to be smaller than *in situ* populations and consequently they are more exposed to genetic drift. On the other hand, *ex situ* live conservation might facilitate proper genetic management since it may be possible to have a high level of control on the entire population in terms of selection and mating of sires and dams. These issues are examined in practice in chapter 8.

We should again underline that *in situ* and *ex situ* techniques are not mutually exclusive and can be complementary in the development of the strategy for specific breeds in specific contexts.

3. Options for utilisation of self-sustaining local breeds

The analysis of the dynamics of the erosion of farm animal genetic resources worldwide will probably reveal a complex and many-faceted picture. It may show cultural, social and food demand changes, transformations of the food production chain, technological

Table 2.2. Conservation techniques and genetic factors.

	Technique		
Factors associated	**cryocon'**	***ex situ* live**	***in situ***
Breed evolution / genetic adaptation	no	poor	yes
Increased knowledge of breed characteristics	poor	poor	yes
Exposure to genetic drift	no	yes	yes

changes, changes in country regulations and importation policies, changes in gross national product (GNP) and marketing activities from multinational breeding companies. The globalisation process affects in various ways the decline of local breeds. In most cases it is likely that these factors conspire to result in a lack in economic profitability of the local breed compared to other breeds or crosses, or to other economic activities in the region. The fall in population size is the first result of these facts often leading to extinction of the local breed. In this respect *in situ* conservation should be triggered when possibilities for breed recovery are still present and when it supports the sustainable use of the resource.

The technique of maintaining endangered breeds in their production environments (*in situ*) covers the widest spectrum of conservation objectives (see Table 2.1). Only continuous utilisation maintains breeds as dynamic entities, adapted to both the needs of the society and the production environment. The key factor for minimising costs for conserving breeds *in situ* is to maintain breeds that retain the potential to be economically self-sustaining. Then, it becomes important to analyse the options we have for an efficient *in situ* conservation of breeds through their utilisation and to make local endangered breeds self-sustainable.

Six general options for in-situ conservation will be considered:

- establishing the economic performance of the breed;
- improving infrastructures and technical assistance;
- genetic improvement;
- optimisation of the production system;
- developing activities to increase the market value of breed products;
- developing incentives.

The large differences among areas of the world, in particular on the basis of gross national product and available technology, would suggest treating groups of countries separately. However, considering the rapid changes we observe in some areas of the world and the fact that there is a continuum of situations rather than discrete groups, we can analyse different options for self-sustaining breeds world-wide. But we will remark upon and indicate opportunities of transfer specific options across countries/areas of the world.

Rural communities and farm animal resources are interdependent and cannot be separated. This criterion is a world-wide rule but it is particularly true in many developing countries with pastoral and smallholder communities. In 2001 an international workshop (Köhler-Rollefson, 2003) underlined the importance of developing "Community-based Management of Animal Genetic Resources". This approach relies on a farm animal genetic resources and ecosystem management where the community is responsible for decisions on defining, prioritising and implementing actions. Some projects are adopting and testing the community-based approach (Köhler-Rollefson, 2003). Focussing on rural communities allows the simultaneous promotion of the development of rural communities and the conservation of animal genetic resources. More generally, an active participation of farmers and all stakeholders, including commercial companies, is important for the success of the options analysed below.

3.1. Establishing the economic performance of the breed

For most of the local breeds we have no reliable data on their performances. Most often:

- performances are estimated on small samples;
- information refers only to phenotypic data, with no estimates of genetic parameters;
- information is not available on fitness traits, such as longevity, fertility, mortality, feed and management requirements characters, which significantly contribute to breed profitability.

In many areas on the world comparisons of performances between crossbred and indigenous breeds have been based on poor experimental designs, which often produce misleading results (FAO, 1998). It is likely that better evaluations of the economic performances of local breeds may change the ranking of local and exotic breeds. It may correct erroneously perceived differences, and may indicate possible strong points of the local breed.

Breed comparisons should first be based on a good assessment of breed performances. When the breeds participate in a national recording scheme for production traits, the biological information for the comparison can be gathered much more accurately. More accurate comparisons require:

- Awareness for interactions between the farm management and the characteristics of the breed, which requires comparisons of breeds in different management systems. For example, the level of inputs made may interact with breed performances.
- Additional trials at experimental stations or at practical farms under controlled conditions. Despite possible high costs, these trials offer the opportunity to compare breeds accurately for input and output factors, which is essential for a proper economic comparison.
- Studies on farms and in experimental stations including the evaluation of purebreds and crossbreds to better understand the potential use of the local breed in different production systems, and to estimate the heterosis involved in crossbred performance.

The relative economic advantage or disadvantage of a breed is a function of the relative prices for the different animal products. A breed, which is not used in high-input, high-output systems, can be profitable in a low-input system through a high feed intake capacity, longevity, fertility, hardiness, quality of the products or a niche market for its products (see Box 2.2). In valuing breeds, besides the traditional products such as milk, meat, fibre, draught, etc., services such as insurance for the future development of animal production, environmental and cultural functions should be taken into account. Assessing the economic values of all these components is fundamental to inform policy decisions, to design economic incentives and to create added value to local breeds. Some research in this respect has been developed in the last years (a Special Issue on Animal Genetic Resources (Anonymous, 2003) of Ecological Economics was dedicated to this aspect). In particular some components/values of farm animal genetic resources are not captured by the market. Specific methods can be used for their economic valuation, such as techniques based on simulating hypothetical markets to estimate the willingness to pay by the society and the willingness to accept by farmers. Reviews on economic valuation of farm animal genetic resources can be found in Drucker *et al.* (2001, 2005), Roosen *et al.* (2005), and a brief overview is presented in chapter 9.

3.2. Improving infrastructures and technical assistance

Most often local breeds produce in areas characterised by specific local socio-economic development. This might be associated with lack of infrastructure and technical assistance, including networks for milk collection and processing, slaughterhouses, networks for commercialisation of products and performance recording. The absence of breeders associations or breeding programmes plays a negative role on breed

Box 2.2. The use of high altitude pastures by the Abondance and Tarentaise cattle: An example of the establishment of the performance of breeds in a specific environment.

The French Northern Alps are famous for their landscapes and opportunities for tourism and sport. Dairy cattle production is the main agricultural activity where milk is processed for cheese production. In particular two cheeses, the Reblochon and the Beaufort, both produced under a Protected Designation of Origin (PDO) in small dairy factories, increased substantially in market share over the last 25 years. Two local cattle breeds, the Abondance and the Tarentaise, play a central role in these cheese productions. Comparisons of these breeds with other French dairy cattle (Verrier *et al.*, 2005) revealed that the Abondance and Tarentaise have: walking capacity with a low impact on dairy production, better resistance to heat, better grazing activity on high altitude pastures under harsh climatic conditions, and better roughage intake, fertility, longevity, somatic cell count, milk protein to fat ratio, milk clotting quality, cheese yield.

sustainability. It is likely that the removal of these constraints will create conditions and opportunities for increasing the economic performance of local breeds.

3.3. Genetic improvement

In general local breeds do not benefit from modern breeding techniques as much as they should. Selection programmes may increase genetic ability for productivity and consequently profitability of local breeds. However, three major considerations have to be forwarded:

- Breeding goals should take into account the conservation values of the breed. Traits proposed for selection should be accurately evaluated for their genetic correlations with those traits that determine the conservation value of the breed, in order to avoid their deterioration. These might include adaptation to a harsh environment or to low-input production systems or traits like longevity, fertility and quality of meat and milk.
- Breeding schemes should be adapted to the farming environment. There is a need for further research on this aspect and implementation of breeding schemes adapted to low/medium input production systems.
- Selection schemes should take into account maintenance of genetic variation within the breed and risks associated with high rates of inbreeding. To reach these goals, a theoretical framework has been developed in the last years and software is available for field use (see chapter 8).

3.4. Optimisation of the production system

In addition to genetic improvement, increasing the economic performance of local breeds might require re-organisation of their production systems, such as seasonal planning of the production, changing age or weight at slaughter or introducing some crossbreeding.

Attention should be given to the conservation of the local breed. Taking the introduction of crossbreeding as an example:

- Breeding schemes should guarantee the maintenance of viable populations of the local breed through a sound pure breeding scheme.
- The breed might be used for the production of commercial crosses with a high performance breed. The commercial crosses might benefit from higher input production systems, while the local breed should be maintained in its original production environment to maintain its adaptation characteristics.

- The use of the local breed as a female population (instead of as male population, which might be more profitable) may be advisable to guarantee the maintenance of a large population of the local genotype adapted to the production environment.
- The use of a high performance breed that will produce crosses that cannot be distinguished from the local breed is not advisable, because of the risk of involuntary introduction of exotic genotypes into the local breed.

3.5. Developing activities to increase the market value of breed products

Successful initiatives were developed in the last years to increase the commercial value of the traditional productions of local breeds. On the contrary, there is still a need for strategies to induce the market recognising not traditional products, such as the cultural and environmental breed services. This paragraph discusses these two aspects.

3.5.1. Links between products and breeds

Generally the control and the enhancement of the quality of agricultural products is a combination of the raw material (meat, milk) and the processing. Many local breeds give products of higher quality with respect to those of commercial breeds that were highly selected for quantitative production. In those countries where the market is ready to recognise the quality of the products of local breeds, the traditional relationship between local breeds and products has been used to diversify products. In this way products of local breeds are sold at a higher price, which improves their profitability. In the areas of the world where food security is given greater consideration, such as most African regions, this approach can rarely be considered. However, an awareness of this option for local breeds is recommended all over the world.

Many successful experiences were developed in the last years supporting the approach of a marketing link between products and local breeds, including dairy and meat products, such as many cheeses and the famous ham from the Iberian pig. Some examples are given in Boxes 2.3, 2.4 and 2.5. The increased interest in Europe for regional food products, including the development of specific organisations such as Slow Food (www.slowfood.com), created favourable conditions for these experiences.

Some general conclusions can be drawn from known experiences:
- The link between product and breed can improve breed's economic profitability.
- Building this link offers several options: e.g. the link can be part of a protected designation of origin (PDO), such as occurs in Europe, or can be used to further differentiate a product within a market already differentiated (e.g. within a PDO).

Box 2.3. Niche products linked to specific breeds: the Parmigiano Reggiano cheese and the Reggiana cattle.

In the 1940's the Italian dairy cattle Reggiana counted more than 40,000 cows, then progressively dropped to a minimum of 500 cows in the early eighties. It was a typical process of displacement of a local breed by the cosmopolitan better promoted and more productive Friesian breed. In 1991 a consortium of breeders started the marketing of a brand of Parmigiano Reggiano made only with milk of the Reggiana breed (the original breed, before the Friesian, was used to produce this cheese). Since its appearance on the market, consumers were ready to pay this branded Parmigiano Reggiano from 30 to almost 100% more with respect to the generic one (Gandini *et al.*, 2007a). The increase of cow population by almost 100% since 1993 (1,250 in 2004) is considered an outcome from this initiative.

Box 2.4. Niche products linked to specific breeds: Skyr, the Iceland dairy cattle product heritage from the Vikings.

Skyr is produced from skimmed milk of Iceland cattle. It is rich in protein and vitamins, with low calories and 18-20% of dry matter. It is served to small children, to schoolchildren at lunch, it is used as dessert and it is a popular 'fast food' in Iceland to day. The industrial production of Skyr started in 1929 and its consumption increased in the last decade. Today the product is exported to the USA as a healthy product from the Icelandic cattle with a specific cultural heritage "Skyr, the national food of Iceland, heritage from the Vikings".

Box 2.5. American Standard Turkeys: an example of using the genetic resource for developing special historic food.

Combined efforts of the American Livestock Breeds Conservancy and Slow Food have led to the resurgence of American Standard Turkeys (Nabhan and Rood, 2004). The turkey was domesticated by the Aztecs over 2000 years ago. Historically, all eight American standard varieties were raised regionally on small family farms. They are excellent foragers, hardy and disease resistant, but they are smaller and slower growing than industrialised stocks. Until the 1940s, the turkey was strictly a seasonal delicacy, synonymous with holiday celebrations. The standard varieties also make superior table birds with their dense but succulent meat and rich complex flavours. In 2001, Slow Food launched an incredibly successful campaign to promote heritage turkeys in restaurant and holiday fare. Nowadays the American Standard Turkey is back on the road to recover thanks to a few breeders, which remained committed to these varieties.

- The overlap of an exotic breed in the farming area of the local breed might hamper the creation of a link between the local breed and the product (e.g. difficulty of separate milk collection).
- In some cases, the link between product and the breed-environment seems to be more appropriate than the link between product and breed.

3.5.2. Ecological and cultural breed products

With particular reference to the European context, we may consider that:

- Before the intensification and industrialisation process in the last decades, livestock farming was closely linked to the use of farmland and in general was extensive. Most of the areas which are recognised nowadays as natural areas are in fact agro-ecosystems created and maintained by farmers and their local breeds. In some cases we might identify a co-evolution process between the breed and the agro-ecosystem. The declines of local breeds and of their production systems are raising concern for the maintenance of these agro-ecosystems and cultural landscapes. Examples include the alpine landscapes characterised by the summer pasture of cattle, sheep and goat herds, the Mediterranean oak forests of the Iberian peninsula (La Dehsa) home to extensive pig farming, the dry pustza grasslands of south-east Hungary and the moors and heaths of north-west Scotland.
- When grazing ceases, bush encroachment follows, which makes it more difficult to use the lands for recreation. Farmers maintain landscapes of great beauty, which are rich in culture. Examples in this respect are the Alpine pastures, which attracts large amounts of tourists in summer.
- The reduction of livestock grazing is known to increase risks associated to natural fires, especially in the Mediterranean regions and to floods in the alpine areas.
- Local breeds have often played a central role, for relatively long periods, in the agriculture tenures and in the social life of rural populations. They are historical witnesses with respect to the rural life.
- Today local breeds are often a reference point of ancient local traditions, such as food, artisan crafts and folklore, and play an important role in the protection of the local cultural heritage, including rural landscapes. In this light, ancient local breeds are vital elements of what we might define as 'cultural networks'.
- Typical products of animal origin, the value of which is today recognised and protected by European directives, originate from specific local breeds, farming methods and areas. These products have become part, over time, of the way of life of rural populations, gastronomic traditions, religious and civic festivals and bear a recognised cultural value.

Based on these and similar considerations, several countries and the European Community developed specific agriculture and environment policies, including subsidy systems directed to the management of rural landscapes and agro-ecosystems. However, subsidies are not expected to be available in the long term. Then, the question arises: is it possible to develop a market value for the ecological and cultural services from local breeds? Recent experiences allow some optimism:

- Some work has been carried out in the last years in Europe (Flamant *et al.*, 1995) in developing cultural tourism associated with the farming culture of local breeds. Cultural tourism has been expanding rapidly over the past two decades and further growth is expected in the future, and could likely intersect with conservation efforts for farm animal genetic resources conservation in some poorer areas of the world.
- In South Europe, cheese producers and breed associations have started to envisage an ecological role for their local breeds. For example, from 1993, in Savoy, herd milk production of Tarantaise and Abondance cattle breeds is limited in order to maintain the optimum stocking rate per hectare. On the Italian Southern side of the Alpes, the production of Fontina cheese implies that milk comes from Valdostana cattle taken to alpine summer pastures.
- In several parts of Europe horses are recognised as preferred help to harvest the wood under rough conditions. This may facilitates the conservation of the original heavy European horses.
- Grazing by domestic ungulates to attain high biodiversity and more complete ecosystems (Box 2.6) is a growing management practice. In UK, since its foundation in 1997, the Grazing Animals Project has promoted and facilitated the use of grazing livestock in management of habitats for conservation (Small, 2004).

Box 2.6. Grazing and ecosystem management.

Studies on breed differences in grazing behaviour and vegetation preferences between a high and a moderate yielding dairy cattle breed in Norway showed that there might be breed differences of importance for the management of semi-natural grasslands (Sæther *et al.* 2006). The breeds were the high yielding, modern dairy cattle breed Norwegian Red (NR) and the moderate yielding, old dairy cattle breed Blacksided Trønder and Nordland Cattle (STN). There was no breed difference in time spent on grazing, but NR breed had a higher demand for nutrient rich fodder and thus preferred to graze the most nutrient rich species compared to STN when grazing on shared, not especially nutrient or species rich, semi-natural grasslands. This difference ought to be taken into consideration since loss of vegetation diversity seems to be smaller when using a moderate yielding breed.

3.6. Incentives

Incentive payments to compensate farmers for the lower profitability of the local breeds compared to substituting these breeds with more profitable exotic breeds have been used in several countries (e.g. Italy) and were adopted by the European Community since 1992.

Economic incentives probably have been effective in many cases to halt the decline of local breeds, however such incentives cannot last forever. In addition, in spite of EU's support, rearing local breeds by farmers often remains unprofitable (Signorello and Pappalardo, 2003).

They also put an upper limit on numbers, since income drops when headage thresholds for subsidy are exceeded. Both in Europe and in other parts of the world it seems worthwhile to investigate the use of incentive measures specific to various situations. Within Europe, for example, the elimination of milk production quotas for endangered breeds could effectively promote their farming. More generally, economic incentives should be used to accelerate the process toward breed sustainability rather than to provide a general economic support.

4. Cryoconservation

Gene banks offer important opportunities for the conservation and utilisation of farm animal genetic resources. In paragraph 2 we saw that cryoconservation allows us to achieve some of the conservation objectives. Combinations of *in situ* and *ex situ* techniques can provide powerful tools for achieving all conservation aims. In chapter 8, for example, it is shown as cryoconserved material can be used to implement efficient management schemes to control genetic drift in live population of small size.

This paragraph analyses, following the Guidelines for the Constitution of National Cryopreservation Programmes for Farm Animals (Hiemstra, 2003) and some recent literature, some technical aspects in creating gene banks. More detailed information on these elements as well as the organisational aspects, legal issues, sanitary requirements, etc., associated to cryobanks are discussed in detail in the Guidelines by FAO (1998) and ERFP (Hiemstra, 2003) and a brief overview can be found in chapter 9.

Sampling of donors animals, type and amount of genetic material to be stored are functions of available funds, local constraints, availability of biological material and, above all, of the aims of storage. All these elements can vary consistently across time and countries. Cryobanking can be based on different aims, including:

- To reconstruct the breed, in case of extinction or loss of a major portion of the population.
- To create new lines/breeds, in case of breed extinction, by combining the stored material with genetic material from other breeds.
- As a back-up, to quickly modify and/or reorient, the evolution or the selection process of populations.
- To support populations conserved *in vivo* in cryo aided live scheme: (1) as a back-up in case genetic problems occur in the living population (e.g. loss of allelic diversity, inbreeding, occurrence of deleterious genetic combinations); (2) to increase effective population size of small populations and reduce genetic drift (see chapter 8).
- As a genetic resource for research.

Below we analyse the general criteria for selecting donor animals, choosing the genetic material to be stored and the specific amounts. Considering the wide spectrum of possible situations, we will not give general figures for the material to be stored and we will provide some explanatory examples for specific cases.

4.1. Which donor animals?

When we consider cryostorage for a given breed, different cryopreservation aims may require storage of different types of genetic variation, which can be obtained by selecting donors with specific criteria:

- Random sampling, in order to store a representative sample of the breed genetic variation. This is probably the most common case.
- Selecting animals carrying specific genotypes / alleles / haplotypes, to modify and reorient the evolution / selection of the population, for gene introgression or to create new lines / breeds. Individuals can be selected on genetic markers, breeding values, phenotype and pedigree information. Box 2.7 reports the example of the UK semen bank within the eradication programme of scrapie from the sheep populations. The French National cryobank (Verrier *et al.*, 2003) recommends the periodic storage of samples of semen from breeds undergoing high intensity selection, and genetic material from old dual-purpose cattle breeds undergoing selection for specialisation in milk or beef production.
- Maximising genetic variation – Specific cases may require to store a sample of the maximal genetic variation of the breed, as in the case of the French National cryobank (Verrier *et al.*, 2003) that aims to store extreme genotypes that might be useful in modifying the evolution of the population. If pedigree information (and/or molecular markers) on the candidate donor animals is available, we can optimise the contribution in number of semen doses and/or embryos of donor animals to the bank as a function of the kinships among them. Than it is possible to select animals

Box 2.7. Building a gene bank to alleviate the risks associated to scrapie eradication.

Scrapie is a transmissible spongiform encephalopathy (TSE) of sheep. Five haplotypes have been observed to segregate at the sheep prion protein (PrP) gene, ARR, AHQ, ARH, ARQ, and VRQ, conferring different degrees of resistance to scrapie, the first and the last haplotypes being associated to highest resistance and highest susceptibility, respectively. Because possible associations with the bovine spongiform encephalopathy (BSE), breeding plans are being adopted in Europe to eradicate the most susceptible haplotypes from sheep populations, and to increase the frequency, eventually up to fixation, of the ARR aplotype conferring the most resistance. However three risks are associated to eradication of scrapie genotypes (Roughsedge *et al.*, 2006). First, there is the possibility that a new TSE will appear to which the currently favoured ARR haplotype may not confer resistance. Second, there is the risk of loss of favourable traits linked to the eradicated genotypes, even though so far associations between PrP variants and production traits have not been identified. The third risk is the potential loss of genetic variation from the populations associated to high selection on PrP genotypes. As a form of insurance against these three risks, a semen bank is currently under construction in the UK, aiming to store the PrP haplotypes expected to be lost in the populations following the scrapie eradication programme. This specific cryopreservation aim requires to take specific decisions in the way to select donors and in the amount of material to be stored. These decisions were taken a posteriori by first simulating different scenarios for the future reintroduction of the eradicated haplotypes in sheep populations. Concerning the selection of donor rams, in order to avoid the third risk of loss of genetic variation from the populations, a sampling strategy was developed (Fernandez *et al.*, 2006) to analyse the optimal contributions (see chapters 5 and 8) of all candidate ram donors, in order to achieve the target frequencies of the removed haplotypes in the cryobank, while maintaining genetic variability in other loci unlinked to those objective of the eradication programme.

in order to minimise the genetic overlapping among the selected candidates. This can be done following Caballero and Toro (2002), or by using the core set method of Eding *et al.* (2002) (see chapters 5 and 8) as:

$$K_a = \Sigma_i \Sigma_j c_i c_j K_{ij},$$

where Σ_i (Σ_j) is summation over all donor candidates; K_a is the average kinship among selected donors, K_{ij} is coefficient of kinship between/among candidate animals i and j; c_i (c_j) is the proportional contribution of animal i to the core set.

The selection of animals that should contribute to the gene bank is done according to the contribution to the core set. If for example an animal shows a contribution equal to zero, it should not be considered as a donor because its genes are already represented by other animals that would be selected as donors to the gene bank. The optimal contribution c_i, that minimises K_a, is explained in chapters 5 and 8. An application of this method for selecting founder animals in aquaculture breeding programmes is given by Hayes *et al.* (2006). Instead of considering the cryopreservation of a single breed, the objective can be to freeze a germplasm pool made of contributions from different populations. In this case, a method for calculating the optimal contributions from each population has been shown to be a valuable technique (Toro *et al.*, 2006; see also chapter 5).

Finally, in general, in some cases it might be necessary through appropriate mating to produce specific donor animals.

4.2. How many donors?

The number of individuals, and the degree of relationship among them, used as donors affects the amount of genetic variation stored in the cryobank. In case of random sampling, using heterozygosity as a parameter of genetic variation, the proportion of breed heterozygosity retained in the bank is, in general, $1 - (1/(2N))$, where N is the number of donor individuals. The use of 25 donors, corresponding to 98% of heterozygosity retained, has been often suggested (e.g. Smith, 1984). When we are interested in capturing allelic diversity and we do not have marker information on the potential donors, the probability to have in storage a specific allele is a function of its frequency (p) in the sampled population and of the number of donor individuals (N), equal to $1 - (1\text{-}p)^{2N}$. Specific objectives may require specific numbers of donors.

4.3. Which type and amount of material to be stored?

In Table 2.3, genetic materials, currently available or in development, for storage are compared for their biological effectiveness to achieve some cryoconservation aims. Costs will be discussed in paragraph 5.

Oocytes differ from embryos, in terms of efficiency to achieve the various aims of cryoconservation, because with oocytes it is still possible to choose the desired mating. Nevertheless, oocytes here are not distinguished from embryos. Embryos can be the first option for breed re-establishment, followed by somatic cells (cloning), assuming this technique will be soon available for this purpose. Semen must be the genetic material of choice for creation of synthetic breeds, for gene introgression and as aid to the genetic management of *in situ* or *ex situ* live programmes. For these purposes embryos and somatic cells are less efficient. The breed can also be re-established by using semen for backcrossing females. The number of generations of upgrading (n), where generation 1 is the F1 cross, and generation n is the n-1 backcross generation, determines the expected proportion (1-

Table 2.3. Conservation aims with different *ex situ* techniques.

	Ex situ technique		
aim	semen	embryos	somatic cells
breed reconstruction	yes* but < 100%	yes	yes*
creation of synthetic breeds	yes	poor	poor
gene introgression	yes	poor	poor
cryo aided live scheme	yes	poor	poor
QTL studies	yes	poor	poor

*Extra-chromosomal DNA can not be recovered, therefore cytoplasmic effects will be lost.

$.5^n$) of genes of the frozen semen present in the last backcross generation. For example, a reconstruction scheme with five generations of upgrading will correspond to an expected recovery of 97% (Hill, 1993) of the original genome in the reconstructed population. The approach with semen has some limitations, including the fact that one can not recover 100% of the genetics of the original breed and, as with somatic cells, cytoplasmic effects will be lost or altered. Gene percentage can be increased marginally by increasing the number of back-cross generations, but the number of doses of semen needed will also increases, in some cases exponentially. In addition, in case of breed reconstruction, when the female reproductive potential is low (e.g. cattle, horse), the number of semen doses required can be very high. To overcome these problems, combinations of semen and embryos can be used instead of only semen or only embryos (Boettcher *et al.*, 2005). In this case breed reconstruction is accomplished by multiplication of a certain number of females from frozen embryos with frozen semen.

Combinations of semen and embryos can be adopted also in breeds of very small size where, due to scarcity of female donors, it might be impossible to obtain the embryos needed with the strategy embryos only. Breed variation could be recovered also using stored semen on a small number of animals kept within an *ex-situ* live scheme. In Box 2.8 it is shown how to compute the amount of genetic material when the cryoconservation target is to store material for future breed reconstruction. Figures given in Box 2.8 are expectations and if numbers are small variances can be high. Simulations to achieve a higher confidence on achieving objectives may be necessary (e.g. Boettcher *et al.*, 2005). A common recommendation is to obtain enough genetic material for two banks. The material can then be stored in separate locations, to minimize risk associated with natural disasters or simple accidents. The availability in the future of technologies such as sexing of embryos and semen and *in vitro* fertilisation at low cost might reduce

Box 2.8. Amount of genetic material to be stored in a gene bank to reconstruct a breed.

This box presents the amount of genetic material to be cryopreserved to reconstruct, in case of breed extinction, a population of 25 females and 25 males of breeding age.
The number of embryos needed for reconstructing 25 females using exclusively embryos can be computed as,

N° of embryos = 25 / (pf × c × sr × sb), (Eq. 2.1)

where: *pf* is the probability that the embryo is female; *c* is the conception rate; *sr* and *sb* respectively the probabilities of survival of the recipient until parturition and of the newborn from birth to breeding age. We assumed not-sexed embryos, thus we will obtain also 25 males.
Breed reconstruction by using only semen is accomplished by creating F1 crosses followed by a series of back-cross generations. The number of semen doses needed for reconstructing 25 females can be computed as (Ollivier and Renard, 1995):

N° of semen doses = d × F × np, (Eq. 2.2)

where: *d* is the number of doses needed per parturition; *F* is the number of females to be inseminated during the reconstruction process to obtain the final 25 females, computed as $25 \times (r + r^2 + ... + r^n)$, where *r* is the inverse of the expected lifetime production of fertile daughters by female assuming on average *np* parturitions. By culling females at a given time, we can set *np* below the species average: in this way reconstruction time decreases, but *r*, *F* and the number of semen doses will increase. Finally, *n* is the number of generations of grading up we decide to use. Because we assumed unsexed semen, at the end we will also obtain 25 males. Please note that the use of sexed semen will substantially reduce the amount of semen to be stored. Some within family selection is advisable during reconstruction and it should be taken into account by setting the parameter *F*.
With the option of storing combinations of semen and embryos, reconstruction starts with a number of females from frozen embryos smaller than 25, followed by their multiplication using semen from the bank until full reconstruction is reached (25 females). In this case the use of Equation 2.2 will overestimate the number of semen doses because longevity of females is not taken into account. The amount of semen can be computed by computer simulation. If we assume unsexed semen, at the end of the reconstruction process we will also obtain 25 males.
During reconstruction using only semen or semen plus embryos, attention should be given to minimise loss of genetic variation due to genetic drift.

▷ ▷ ▷

Example (goat breed):
A) Storage of only embryos (assume: pf =.5; c =.4; sr =.9; sb =.8): 174 embryos to be stored.
B) Storage of only semen (assume: n = 5; d= 2; r =.67 (np was set at 3 to avoid reconstruction time above 10 years), than F = 45): 270 semen doses to be stored.
C) Storage of embryos plus semen; some results from Boettcher *et al.*, (2005) (in this case, no limitations on number of parturitions were used: np = 4; r =.5): 43 embryos and 65 semen doses or 108 embryos and 45 semen doses.
Time for reconstruction will be: only embryos = 2 years; only semen = 9.4 years; embryos plus semen = 5.4 and 3.2 years, respectively for the two cases reported above. Amounts of embryos and semen computed above should be doubled to create two storage sites.

the amount of the material to be stored. Extraction of semen from the epididymus of slaughtered animals offers additional opportunities in constructing semen banks (Gandini *et al.*, 2007b).

Efficiency of freezing and reproduction techniques is progressively improving, but some differences still exist among species. Table 2.4 shows the state of the art of cryopreservation techniques, which includes the efficiency at freezing and after freezing.

Since the sheep *Dolly* was recreated from udder somatic cells, by cloning methods we can re-establish animals from their somatic cells. Other species were cloned by this method as e.g. horse, cattle, pigs, dog, cat, and by cross-species nuclear transfer. Storage of somatic

Table 2.4. State of the art of cryopreservation, by species: + = routine technique available; 0 = positive research results, * = some research hypotheses; – = not feasible in the present state of art; ? = technique unknown (from Hiemstra, 2003, modified).

Species	Semen	Oocytes	Embryos	Somatic cells
Cattle	+	+	+	0
Sheep	+	0	+	0
Goat	+	*	+	0
Horse	+	0	0	0
Pig	+	0	0	0
Rabbit	+	?	+	0
Chicken	+	-	-	0
Fish - some species	+	*	*	*
Dog	+	?	?	0
Cat	0	0	0	0

cells is cheap (Groeneveld, 2005), but up to now we can not use it as a regular method for re-establishing animals or breeds. Considering future developments of scientific knowledge and the relatively low costs, somatic cell freezing could be considered. Oocyte freezing is possible in cattle, but efficiency of *in vitro* development after fertilisation and survival rate after freezing are still low. In conclusion the available tools are semen and embryos storage, although somatic cells and oocytes freezing could be used if no alternatives are possible, and as additional tools. A more comprehensive analysis of the state of the art of cryopreservation technologies can be found in Hiemstra (2003).

5. Costs

Funds for financing conservation of farm animal genetic resources are rather limited. Costs are expected to differ significantly among breeds, contexts, countries and regions. Little field data are available in the literature. Costs of *in situ*, *ex situ* and combined *in situ* and *ex situ* schemes have been analysed for conservation programmes of African cattle on a time horizon of 50 years (Reist-Marti, 2006). The development of conservation policies and programmes requires the analysis of the associated costs. A model for optimal allocation of conservation funds was proposed by Simianer *et al.* (2003). Comparing costs of *in situ* and *ex situ* techniques needs:

- To define the conservation time horizon. With cryoconservation, costs of maintenance are a function of the number of years before use. A possible advantage of cryoconservation over *in situ* conservation can be expressed by the number of years of conservation above which it becomes cheaper than keeping live animals.
- To define whether costs for restoring the frozen material to the form that can be used should be included. Cryoconservation with semen can be consistently more costly when breed re-establishment costs are considered (Lomker and Simon, 1994).

These comparisons therefore require assumptions and information specific to the breed/context. Due to the scarce literature, only some considerations can be done on the three techniques:

- *In situ*: The approach proposed in this chapter is the conservation of farm animal genetic resources though their sustainable utilisation. It follows that: (1) incentive payments to farmers should cover the gap in economic return between the endangered breed and the average commercial breed. Breed comparisons are useful to determine the amount of the incentive payment; (2) the length of economic incentive payments depends on the period necessary to take the breed towards self-sustainability; (3) taking the breed towards self-sustainability may imply costs for technical assistance, for the development of breeder associations, for performances recording and breeding schemes, and costs to identify, qualify and market products linked to the breed, including cultural and ecological services. It should be underlined

that, when self-sustainability is reached, conservation costs become zero. Costs of cryo aided live schemes should be accounted for, where applicable.
- *Ex situ* live: Costs of *ex situ* live conservation are equal to the difference between the profit from farming the average commercial breed and the endangered breed, similarly to *in situ* at the start. Market strategies promoting tourism (herds kept in natural protected areas, in show-farms) and high quality products can be used to increase profitability. Costs of cryo aided live schemes should be accounted for, where applicable.
- *Ex situ – Cryoconservation*: Total costs of cryoconservation include collection and storage costs, and, if applicable, costs for long term maintenance of the material and for breed reconstruction. Cost of collection and storage are functions of the amount of material needed and unitary costs. The latter will vary across countries and areas. In a simulation work, high variation of costs have been observed among the main livestock species, banking strategies - storing only semen, embryos, or semen and embryos combinations - and scenarios - presence or absence of a market for the breed semen, standard semen collection or extraction of semen from the epididymus, presence or absence of costs to buy donor animals - (Gandini *et al.*, 2007b). In general storing only embryos is more expensive than storing embryos and semen and only semen, because the higher costs in collecting embryos in most of the species. However, in some cases costs for storing embryos plus semen are not significantly different from those for storing only semen. It general it seems that no single recipe for gene banking is universally superior and tools are needed to easily estimate costs as a function of specific cryo aims, breeds and contexts.

6. Making the decision

Often, because financial and human resources are limited, breeds cannot be given the same priority for conservation. Chapter 6 discusses criteria and methods for selecting breeds for conservation. Breeds will be selected on the basis of a general policy, such as conservation of genetic diversity for future uses and/or maintenance of rural cultural diversity and/or development of rural areas, and by using tools to reach policy objectives while minimising costs in terms of number of breeds. The choice may vary among countries, because the different interest and priorities of Governments and human societies. At the same time of selecting, the question arises: which *in situ* or *ex situ* technique or which *in situ* and *ex situ* combination should be used?

In paragraph 2, the capability of each technique to reach the conservation objectives was analysed. *In situ* and *ex situ* techniques were discussed more in detail in paragraphs 3 and 4., respectively. A framework for analysing conservation costs with the different techniques was presented in paragraph 5. Here we present a method to choose the most appropriate conservation technique, based on the following 5 steps (see Figure 2.1):

1. **Which conservation objectives apply to the breed?** An accurate evaluation of the conservation values of the breed is expected to be done in selecting breeds for conservation. Remember that to analyse genetic variation within and between breeds we have objective criteria, but we do not yet have standardised tools to evaluate other breed values such as the ecological and socio-economic services of the breed.

2. **Rank the techniques for their efficacy to reach the conservation objectives.** Not all techniques address the same conservation objectives with equal effectiveness. Following Table 2.1, rank techniques (*in situ, ex situ* live, cryconservation), or technique combinations, as a function of their efficacy to reach the conservation objectives previously identified (conservation policy). If sustainable utilisation of rural areas is an objective, maintenance of the breed within its production system (*in situ*), and in some cases *ex situ* live, is the only technique available. Cryoconservation can be used in addition to *in situ* and *ex situ* live to reduce the risk of losing a breed. In the ranking procedure, factors associated with techniques (see Table 2.2), such as opportunity for breed evolution/adaptation and for obtaining a better knowledge of breed characteristics, should be considered.

3. **Rank techniques for risk of failure.** Analyse techniques or combinations of techniques on the basis of the risk of failure and exclude those with a non-acceptable level of risk. As an example, with *in-situ* the risk of failure can be a function of, among others, the:
   - possibility of controlling genetic drift (e.g. availability of skilled technicians and farming structure for appropriate genetic management);
   - possibility of removing factors that in the past limited or impaired conservation success;
   - possibility of having economically self-sustainable populations;
   - probability of socio-economical and political instability;
   - probability of epidemics, etc.

4. **Rank techniques for costs.** For those techniques or combinations of techniques that guarantee acceptable risks of failure, costs should be evaluated. Rank all techniques on a cost basis.

5. **Choose the technique.** Consider the different rankings for (1) efficacy in achieving the conservation aims; (2) risk of failure; and (3) costs. The weighing of each of these factors will depend upon interests and priorities, resources available, strategy preferences.

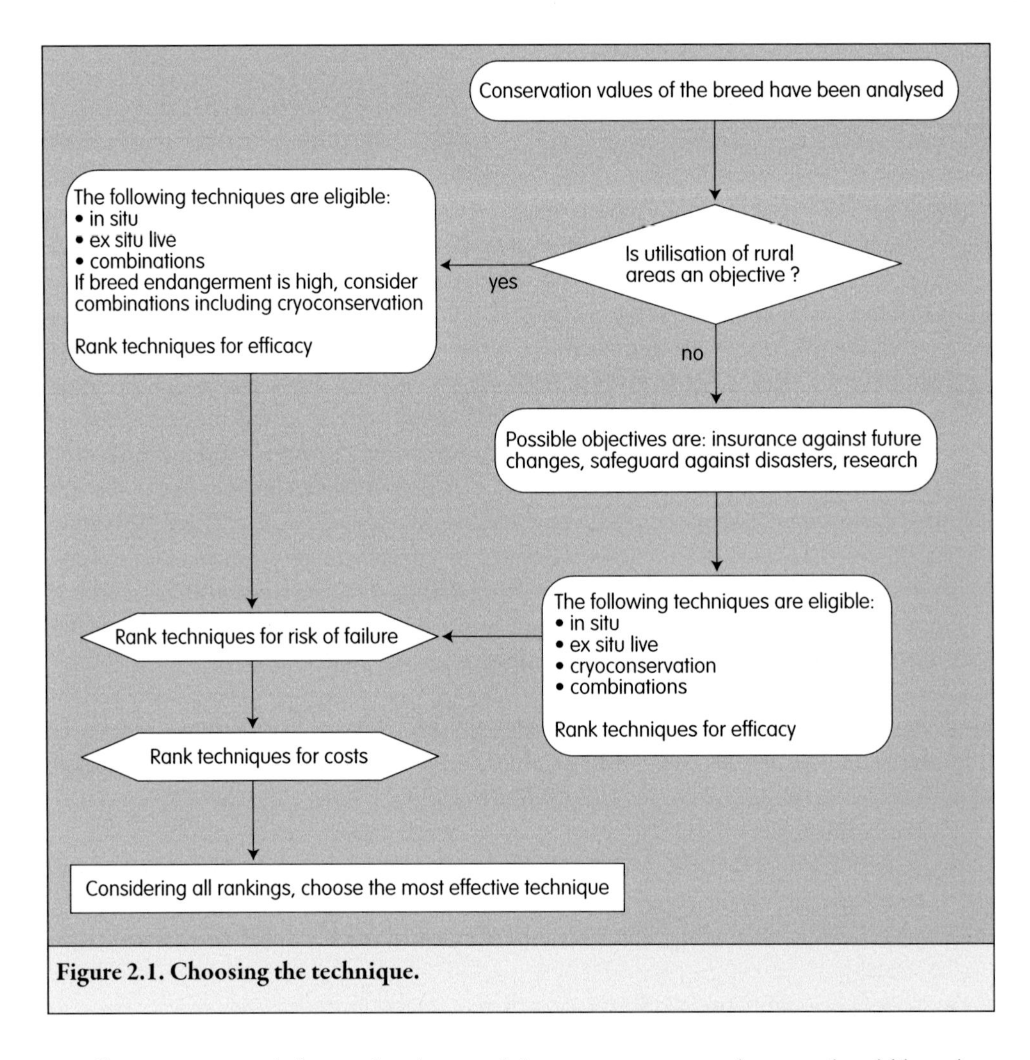

Figure 2.1. Choosing the technique.

Finally, as mentioned above, the choice of the conservation technique should be taken at the same moment as the selection of the breeds and taking into account that these two different processes might influence each other.

References

Anonymous, 2003. Special Issue on Animal Genetic Resources, 2003. Ecological Economics, 45 (3).

Boettcher, P.J., A. Stella, F. Pizzi and G. Gandini, 2005. The combined use of embryos and semen for cryogenic conservation of farm mammal genetic resources. Genetics Selection Evolution 37: 657-675.

Caballero, A. and M.A. Toro, 2002. Analysis of genetic diversity for the management of conserved subdivided populations. Conservation Genetics 3: 289-299.

Drucker, A.G., V. Gomez and S. Anderson, 2001. The economic valuation of farm animal genetic resources: a survey of available methods. Ecological Economics 36: 1-18.

Drucker, A.G., M. Smale and P. Zambrano, 2005. Valuation and sustainable management of crop and livestock biodiversity: a review of applied economics literature. CGIAR, IFPRI, IPGRI, ILRI.

Eding, H., R.P.M.A. Crooijmans, M.A.M. Groenen and T.H.E. Meuwissen, 2002 Assessing the contribution of breeds to genetic diversity in conservation schemes, Genetics Selection Evolution 34: 613–633.

FAO, 1998. Secondary Guidelines for Development of National Farm Animal Genetic Resources Management Plans: Management of Small Populations at Risk. FAO, Rome, Italy.

Fernandez, J., Roughsedge T., J.A. Woolliams and B. Villanueva, 2006. Optimization of the sampling strategy for establishing a gene bank: storing PrP alleles following a scrapie eradication plan as a case study. Animal Science 82: 813-821.

Flamant, J.C., A.V. Portugal, J.P. Costa, A.F. Nunes and J. Boyazoglu (eds.), 1995. Animal Production and rural tourism in Mediterranean Regions. EAAP Publication No 74. Wageningen Pers, Wageningen.

Gandini, G.C. and E. Villa, 2003. Analysis of the cultural value of local livestock breeds: a methodology. Journal of Animal Breeding and Genetics 120: 1-11.

Gandini, G, C. Maltecca, F. Pizzi, A. Bagnato and R. Rizzi, 2007a. Comparing local and commercial breeds on functional traits and profitability: the case of Reggiana dairy cattle. Journal of Dairy Science (in press).

Gandini, G., F. Pizzi, A. Stella and P.J. Boettcher, 2007b. The costs of cryogenic conservation of mammalian livestock genetic resources. Genetics Selection Evolution (in press).

Groeneveld, E., 2005. A world wide emergency program for the creation of national genebanks of endangered breeds in animal agriculture, AGRI 36: 1-6.

Hayes, B., J. He, T. Moen and J. Bennewitz, 2006. Use of molecular markers to maximise diversity of founder populations for aquaculture breeding programs. Aquaculture 255: 573-578.

Hiemstra, S.J. (ed.), 2003. Guidelines for the Constitution of National Cryopreservation Programmes for Farm Animals. Publication No. 1 of the European Regional Focal Point on Animal Genetic Resources.

Hill, W.G., 1993. Variation in genetic composition in backcrossing programs. Journal of Heredity 84: 212-213.

Köhler-Rollefson, I., 2003. Community-based management of farm animal genetic resources. In: Farm Animal Genetic Resources - Safeguarding National Assets for Food Security and Trade. Proceedings of the workshop held in Mbabane, Swaziland 7-11 May 2001. GTZ, CTA, FAO, Rome, Italy.

Lomker, R. and D.L. Simon, 1994. Costs of and inbreeding in conservation strategies of endangered breeds of cattle. Proceedings of the 5th World Congress on Genetics Applied to Livestock Production, Guelph, 21: 393-396.

Nabhan, G.P. and A. Rood, 2004. Renewing America's Food Traditions. The Center for Sustainable Environments at Northern Arizona University, Flagstaff Arizona.

Ollivier, L. and J.P. Renard, 1995. The costs of cryopreservation of animal genetic resources. Proceedings 46th Annual Meeting of EAAP, Prague.

Reist-Marti, S.B., A. Abdulai, and H. Simianer, 2006. Optimum allocation of conservation funds and choice of conservation programs for a set of African cattle breeds. Genetics Selection Evolution 38: 99-126.

Roosen, J., A. Fadlaoui and M.Bertaglia, 2005. Economic evaluation for conservation of farm animal genetic resources. Journal of Animal Breeding and Genetics 122: 217-228.

Roughsedge, T., B. Villanueva and J.A. Woolliams, 2006. Determining the relationship between restorative potential and size of a gene bank to alleviate the risks inherent in a scrapie eradication breeding programme. Livestock Science 100: 231-241.

Sæther, N.H., H. Sickel, A. Norderhaug, M Sickel and O. Vangen, 2006. Plant and vegetation preferences for a high and a moderate yielding Norwegian dairy cattle breed grazing semi-natural mountain pastures. Animal Research 55: 367-387.

Signorello, G. and G. Pappalardo, 2003, Domestic animal biodiversity conservation: a case study of rural development plans in the European Union. Ecological Economics 45: 487-499.

Simianer, H., S.B. Reist-Marti, J. Gibson, O. Hanotte, and J.E.O. Rege, 2003. An approach to the optimal allocation of conservation funds to minimise loss of genetic diversity between livestock breeds. Ecological Economics 45: 377-392.

Small R.W., 2004. The role of rare and traditional breeds in conservation: the Grazing Animals Project. In: Farm Animal Genetic Resources (BSAS Publication n° 30), G. Simm, B. Villanueva, K.D. Sinclair and S. Townsend (Eds.), Nottingham University Press, Nottingham, pp. 263-280.

Smith C., 1984. Estimated costs of genetic conservation in farm animals. In: FAO Animal Production and Health Paper 44/1. FAO, Rome, Italy, pp. 21-30.

Toro M.A., J. Fernandez and A. Caballero, 2006. Scientific basis for policies in conservation of farm animal genetic resources. Proceedings 8th World Congress on Genetics Applied to Livestock Production, CD-ROM Communication No. 33-05.

Verrier E., C. Danchin-Burge, S. Moureaux, L. Ollivier, M. Tixier-Boichard, M.J. Maignel, J.P. Bidanel and F. Clement, 2003. What should be preserved: genetic goals and collection protocols for the French National Cryobank. In: Workshop on Cryopreservation of Animal Genetic Resources in Europe, D. Planchenault (ed.), Paris 2003.

Verrier E., M. Tixier-Boichard, R. Bernigaud and M. Naves, 2005. Conservation and value of local livestock breeds: usefulness of niche products and/or adaptation to specific environments. AGRI 36: 21-31.

Chapter 3. What is genetic diversity?

John Woolliams[1,2] and Miguel Toro[3]
[1]Roslin Institute, Roslin, Midlothian EH25 9PS, United Kingdom
[2]Department of Animal and Aquacultural Sciences, Norwegian University of Life Science, Box 1432, Ås, Norway
[3]Department of Animal Breeding, National Agriculture and Food Research Institute, Carretera La Coruna km7, 28040 Madrid, Spain

Questions that will be answered in this chapter:

- *What is meant by diversity?*
- *In what ways can diversity be quantified?*
- *To what extent are different measures of genetic diversity describing the same phenomenon?*
- *How can we measure change in genetic diversity?*

Summary

This chapter surveys the different ways in which the diversity we observe within a species can be quantified. It will consider: (1) how the diversity we observe in the population of a species can be divided into differences between breeds, or more generally sub-populations, and differences within these breeds; (2) how diversity within breeds can be described by examining the pedigrees of individuals, or examining the sequences of DNA carried by individuals; (3) how a number of different approaches have been used to measure the genetic diversity; and (4) how many of these measures may be related to the concepts of inbreeding.

1. Introduction

Diversity, in its most commonly recognised sense, is the observation of different forms and functions between species. However this definition is too narrow and fails to recognise that individuals within a species also differ in many characteristics (or phenotypes); these phenotypes may be qualitative, such as coat colour or pattern, or quantitative, such as height and weight. In domestic animals particular phenotypes will have evolved because of their utility to particular human populations. The most useful concept in quantifying these differences is the fundamental statistical concept of variance since it can be decomposed into sub components, and in particular, if we have a sum of independent variables each with some variance, then the variance in the sum of the variables is equal to the sum of the variances. This property will be used in what follows at several points.

We may define the total diversity in a trait within a species as the variance of the phenotypes. Of course, not all the variance we observe is genetic in origin. For example whilst we may expect a considerable similarity between identical twins, they will not be totally identical since differences will arise through the environmental impact of life-history events from conception onwards. In the study of genetics it is widely assumed as a first and useful approximation that we can decompose the phenotype into two independent components one due to genetic effects and one due to environmental effects i.e. $P = G + E$, where P has total phenotypic variance σ_P^2, G has total genetic variance σ_G^2, and E has total environmental variance σ_E^2. Using the decomposition property of variance for independent effects, described above, we find $\sigma_P^2 = \sigma_G^2 + \sigma_E^2$, and so the total variance of the phenotypes that is observed can be uniquely decomposed into a genetic and environmental component. The fraction of the total phenotypic variance that is of genetic origin is called the *broad sense heritability*, denoted H^2. Therefore by conducting experiments to estimate H^2 the total genetic variance can be calculated in the population.

However experiments to estimate H^2 are surprisingly difficult since they involve the tracking of individuals with identical genotypes, such as identical twins or clones, and these can be hard to find. It is therefore common to focus on a component of the genetic variance, termed the additive genetic variance (σ_A^2). This component has a special utility as it forms that part of the genetic variance that can be used to create changes in the population mean by selection, and is the variance of the breeding values in the population. The ratio of σ_A^2 with the total phenotypic variance is called the *narrow-sense heritability*, denoted h^2, and unless otherwise stated in the literature, particularly in animal breeding literature, the terms 'heritability' and 'h^2' should be interpreted as denoting the narrow-sense heritability (as in this book). Falconer and Mackay (1996) give more details on defining breeding values, heritabilities and their derivation. It is relatively easy to obtain information on breeding values from experimental or field data and so the genetic variance in the population is often estimated simply by σ_A^2. Note that $\sigma_A^2 \leq \sigma_G^2$ and $h^2 \leq H^2$.

The use of variance to summarise diversity is most obviously applicable in the case of continuous traits such as height and weight. However it can also be easily applied to qualitative traits where there are two outcomes e.g. say horns or no horns, since the two classes can be transformed into 0's and 1's and variances calculated using these numbers. More difficult are qualitative traits with more than two classes: where these form an ordered progression, such as categories defined as 'poor', 'average' and 'good', then transformation to a numerical scale is natural and allows a meaningful variance to be calculated; however where they do not form such a progression, as in red, black or white coat colour, the problem has further difficulties although the concept of variance remains important. One perspective on this is that as we learn more about the loci controlling such characters, and the segregating alleles that prompt the qualitative differences, so we can quantify diversity from the variances and covariances of the allele frequencies.

2. Sub-populations and evolutionary forces

Observing sheep in Scotland it would be easy to assume that all sheep have black on their face and long coarse wool, and come to the conclusion that there is little diversity in these traits. This conclusion would be wrong because in other regions of the world, e.g. Australia, white-faced sheep with fine wool predominate! Livestock populations have a considerable number of sub-populations, with these sub-populations formed to varying degrees by geographical isolation, selection by their human keepers and other evolutionary forces. These sub-populations may loosely be termed breeds (Box 3.1) and selection has focused upon their exterior appearance and on performance in important economic traits with different groups of farmers demanding different qualities from their livestock. The development of breeds will be covered in detail in chapter 4, but for the purpose of this chapter it is important to recognise that such sub-populations exist. In what follows sub-populations will be termed 'breeds'.

The impact of the existence of breeds on diversity is that we can further subdivide the observed breeding value for an individual as $A = A_B + A_W$ where A_B is the mean of the individual's breed and is the same for each member of the breed, and A_W is the deviation of that individual from the mean of its breed. Since the breeds vary in mean, these means will have their own variance σ_B^2, and the deviations of individuals from their breed means will also have a variance σ_W^2. Since the mean and the deviation from the mean is a decomposition of the phenotype into two terms that are statistically independent (as described in paragraph 1) we have $\sigma_A^2 = \sigma_B^2 + \sigma_W^2$. Consequently we can see that the genetic diversity of a species may be subdivided into the genetic component between breeds and a genetic component within breeds. The ratio $\sigma_B^2/(\sigma_B^2+\sigma_W^2)$ describes the importance of breed variation to the total genetic variation in the species, and will certainly vary from trait to trait: as the value increases from 0 to 1 so the existence of different breeds is increasingly important to the maintenance of diversity in the species.

Box 3.1. What is a breed?

A simple question but difficult to answer, and the following are published definitions from a variety of groups, each relevant and pertinent to their stakeholders:

1. "*Animals that, through selection and breeding, have come to resemble one another and pass those traits uniformly to their offspring.*" (http://www.ansi.okstate.edu/breeds/, 28/09/2006).
2. "*A breed is a group of domestic cats (subspecies felis catus) that the governing body of CFA has agreed to recognise as such. A breed must have distinguishing features that set it apart from all other breeds.*" (Cat Fanciers Association, http://www.cfa.org/breeds/breed-definition.html, 28/09/2006).
3. "*A race or variety of men or other animals (or of plants), perpetuating its special or distinctive characteristics by inheritance.*" (http://www.biology-online.org/dictionary/Breeds, 28/09/2006).
4. "*Race, stock; strain; a line of descendants perpetuating particular hereditary qualities.*" (Oxford English Dictionary, 1959).
5. "*Either a sub-specific group of domestic livestock with definable and identifiable external characteristics that enable it to be separated by visual appraisal from other similarly defined groups within the same species, or a group for which geographical and/or cultural separation from phenotypically separate groups has led to acceptance of its separate identity.*" (FAO World Watch List, 3rd Edition).
6. "*A breed is a group of domestic animals, termed such by common consent of the breeders... a term which arose among breeders of livestock, created one might say, for their own use, and no one is warranted in assigning to this word a scientific definition and in calling the breeders wrong when they deviate from the formulated definition. It is their word and the breeders' common usage is what we must accept as the correct definition.*" ('The Genetics of Populations'; Lush, 1994).
7. "*A breed is a breed if enough people say it is.*" (K. Hammond, personal communication).

Continuing definition (5), FAO argue that breed is very often a cultural term and should be respected as such, a perspective clearly articulated in definition (6), and succinctly summarised in (7). This is acknowledged, but in the following chapters where the nature and use of diversity is being explored, the concept of resemblance through common hereditary descent is a useful addition to the definition of a breed.

The magnitude and importance of this variation between breeds will be discussed elsewhere (chapters 4, 5 and 6), however $\sigma_B^2/(\sigma_B^2+\sigma_W^2)$ is a key statistic in determining the likely importance of conserving breeds (Box 3.2). Globally there is a large decrease in the number of breeds, with breeds becoming extinct or critically endangered through substitution by other breeds or indiscriminate crossing. These substitutions are generally driven by current economic and market considerations, and generally involve the substitution of breeds that survive well on low to medium inputs by breeds with high inputs and high output. Therefore the loss of breeds is highly selective and results

Box 3.2. The importance of breed variation to diversity in performance.

The variation between breeds in performance has become an important parameter in the strategic planning of livestock development. For a particular trait σ_B^2 gives an indication of how much progress in a trait may be obtained by selection among breeds, and embarking upon development strategies that will ultimately lead to breed substitution. The observation of substantial breed differences has led to the concentration of breed development on fewer and fewer breeds, with increasing number of breeds being considered unprofitable and consequently at risk of extinction. However due to broadly adverse correlations between productivity and adaptive fitness, observed empirically, this strategy risks losing variation between breeds in traits not properly valued by current markets but of potential importance to the sustainability of rural communities in the future. Therefore the key question is what proportion of total genetic variation for quantitative traits lies between breeds? If this parameter is typically small then it may be expected that within-breed variation can be utilised in selection programmes to overcome weaknesses as they arise; if the parameter is large then it may be overly optimistic to expect an adequate selection response within breeds.

Answers to this question are scarce, and have tended to be concerned with growth, efficiency and carcass characteristics: the multibreed trial conducted with 14 British breeds of cattle (Thiessen *et al.*, 1984, 1985), and other work conducted in Clay Centre, USA with Angus or Hereford cows mated to sires from 17 breeds including Sahiwal and Brahman. These studies examined the range of breeds kept in the same environment, although the environments were experimental and limited in the range of environmental stressors.

Thiessen *et al.* (1985) report that the fractions $\sigma_B^2/(\sigma_B^2+\sigma_W^2)$ in food conversion efficiency during growth and relative growth rate were estimated to be ~0.25 and ~0.33 respectively (a ratio they denote g_1^2). The work at Clay Centre was less constrained, including a wider group of breeds and traits such as calving ease, which may be more closely related to fitness. The extent of breed variation as measured by $\sigma_B^2/(\sigma_B^2+\sigma_W^2)$ in the Clay Centre studies was somewhat larger, well in excess of 0.50 for weight traits (Jenkins *et al.*, 1991), calving traits and survival to weaning (Cundiff *et al.*, 1986). A smaller range of breeds was also examined for carcass traits and palatability (Wheeler *et al.*, 1996) where the range of breed means indicated $\sigma_B^2/(\sigma_B^2+\sigma_W^2)$ remained substantial.

Thiessen *et al.* (1985) also compute an alternative value, $g_2^2 = \sigma_B^2/(\sigma_B^2+\frac{1}{4}\sigma_W^2)$, which they justify as the fraction of 'immediately selectable genetic variation' that lies between breeds. Their justification for this is that the genetic variation within breeds requires to be identified, whilst the breed variation is readily identified; therefore they consider the importance of breed variation as a component of the variance of EBVs among all possible newborn purebred female replacements from all breeds sired by males with high accuracy EBVs. g_2^2 was ~0.57 and ~0.66 for growth rate and relative growth rate. Whilst g_2^2 can appear as a somewhat artificial, it is worth noting that it is the value of g_2^2 for high efficiency that is driving the breed loss that is currently observed!

In conclusion the range of values for $\sigma_B^2/(\sigma_B^2+\sigma_W^2)$ remains poorly documented but provide justification for the broad statement that breed variation accounts for approximately ▹▹▹

half the total genetic variation. In the absence of information to the contrary it is reasonable to assume that this fraction will be applicable to a broad range of traits, including fitness for production in low- to medium-input environments with their associated stressors. If crises were to occur that required livestock production to adapt quickly to new challenges then it is the value of g_2^2 that will be important, which will be > $\sigma_B^2/(\sigma_B^2+\sigma_W^2)$. Therefore conserving breeds with diversity of characteristics is a rational and important strategic response to the environmental uncertainties of today.

in lower σ_B^2, i.e. a potential reduction in loss of genetic variation, particularly for fitness in environments with low to medium inputs.

The differences between breeds will have developed through a combination of four evolutionary forces: genetic drift, migration, selection and mutation (Falconer and Mackay, 1996). *Genetic drift* is a term for the random fluctuations of allele frequencies due to random sampling processes involved when genes are passed from parent to offspring, and is one of the phenomena linked to inbreeding. Over time genetic drift will lead to increasing genetic differences between two breeds drawn from the same population and then maintained in isolation. The *migration* of individuals moving from one breed to another, acts against inbreeding, since it lessens the genetic differences that exist between the breeds, and increases the variation within the recipient breed. If *selection* occurs, carriers of favourable alleles have a selective advantage in the next generation, and the scale of differences between two breeds will not necessarily reflect the degree of isolation. Selection may favour convergence or divergence depending on the selection taking place in each breed. In livestock, selection can be both artificial and natural; for example, natural selection will have played an important role in improving adaptive fitness for particular breeds kept over many generations in environments with specific challenges e.g. periodic droughts. In general, *mutation* in the genome increases the genetic differentiation between breeds and creates genetic diversity. However, mutation occurs with a low frequency (chapter 8) and, in the absence of selection, the influence of mutation becomes measurable only over a relatively large number of generations. However at some point in the past, mutation has been responsible for creating the polymorphisms that lie at the heart of all genetic diversity.

3. The use of pedigree for measuring diversity

We can estimate the degree of diversity between breeds by simple, often costly, experiments in which animals of different breeds are kept together in the same environment. Providing (1) the numbers of animals per breed are sufficiently large, so that the errors in estimating the breed mean are negligible compared to the scale of differences between breeds, and (2) the breeds are a fully representative sample of the breeds available, the variance σ_B^2 can be derived from the breed means. Which environment for testing and how different the answers would be in other environments are important research questions that have global implications for agriculture, the environment and conservation. Given this uncertainty, it is important that testing environments are directly relevant to the intended environment for implementation.

Quantifying the amount of genetic variation in a trait within a breed is more difficult and involves associating known genetic similarities between individuals with similarities in phenotypes. The expectation is that more related individuals will resemble each other more closely than two individuals picked at random from the breed. Falconer and Mackay (1996) show how the relationship between individuals can be related to the magnitude of the covariance in their performance, and how this allows the estimation of the heritability, h^2. A major source of the reliable information on the kinship is the pedigree, i.e. a record of sire and dam for each individual, accumulated over generations. In practice, the most informative relationships are often paternal half-sibs (individuals sharing the same sire) since the high reproductive rate of males in livestock species makes them relatively abundant, and they have a covariance ($\frac{1}{4}\sigma_A^2$) that is readily interpretable. In the absence of detailed information on DNA from individual animals, which will continue to be the case for most populations for some time into the future, there is a need to identify these relationships through observing and recording the pedigrees of animals, at least in sufficient detail to identify sires.

4. The impact of DNA information

The last decade has seen the cost of genotype information reduced by orders of magnitude, making such information much more affordable for science and for commercial applications. This is opening up new opportunities for evaluating diversity. A number of different marker types have been used in scientific studies and their popularity has changed with advances in technology. Box 3.3 provides a short review of important properties for markers and how well the different types match up to these properties.

Informative DNA markers can therefore help the measurement of diversity as described in paragraph 3 in two ways. The first way is to overcome the problem that in some species it may be impossible or very costly to observe pedigree directly, e.g. in many fish species; and by genotyping a small number of markers (say 5 to 10), chosen to be informative, on all offspring and all possible parents then it is possible to identify the sires and dams of almost all the offspring. The second involves the extensive genotyping across all chromosomes of genome in order to estimate the actual proportion of DNA shared by sibs or other relatives more precisely than simply using the expectation that

Box 3.3. The attributes and advantages of types of DNA markers.

Marker Attributes: The following desirable attributes for a marker type can be identified.

- *Widely distributed.* Wide distribution of markers throughout the genome allows the mapping of the whole genome and tracking of gene flow in populations, although maybe with only low resolution.
- *Locally dense.* Ability to find many markers within small genomic regions for the purpose of fine mapping.
- *Ability to localise.* The marker can be placed at a physical location in the genome.
- *Highly polymorphic.* The utility of a marker depends on its ability to distinguish between segments of homologous chromosomes. This will depend on the frequency of heterozygosity, which increases with the number of alleles at the marker locus and the more equal they are in frequency. The amount of information contained by a marker is often defined by its information content (Lynch and Walsh, 1998).
- *Co-dominant.* Ideally, both alleles at a marker locus can be distinguished. For some types of marker the individuals carrying 1 or 2 copies of an allele cannot be distinguished.
- *Low Mutation Rate.* For many uses of marker information the long-term stability of the marker over generations is important for inferences about identity by descent (IBD). The higher the mutation rate, the less certain the inferences become. It is often desirable for the marker locus to have a mutation activity that is representative of coding or regulatory sequences.
- *High Throughput.* Determined by a number of elements, including the amenability to PCR, which allows more genotypes to be obtained from the same quantity of DNA, ability to automate and multiplex assay procedures, and speed of assay.
- *Low Technical Cost.* Expressed per genotype.
- *Repeatable.* Assays should be highly repeatable, both between assays within laboratories and between laboratories.

Types of Marker

- *Mini-satellites (MiniS).* A sequence of DNA base pairs, typically of containing 10's of base pairs, repeated a variable number of times.
- *DNA Fingerprints.* Classically a multiple array of mini-satellites.
- *Restriction Fragment Length Polymorphism (RFLP).* An early bi-allelic marker type based on recognition sites for restriction enzymes.
- *Randomly Amplified Polymorphic DNA (RAPD).* Markers formed from an arbitrary set of PCR primers, resulting in a random set of amplified segments.
- *Micro-satellite (MicroS).* Based upon sites in which the same short sequence is repeated multiple times.
- *AFLP.* AFLPs are a multiple array of RFLPs displayed in a single gel.
- *Single Nucleotide Polymorphism (SNP).* Point mutation in the genome sequence, predominantly bi-allelic, but feasible to have 4 alleles, with each of the 4 nucleotide bases appearing in the same location.

▷▷▷

Table 3.1. A general guide to the attributes of different marker types.

Attribute	MiniS	Fingerprint	RFLP	RAPD	MicroS	AFLP	SNP
Widely distributed	Moderate	Moderate	Very good	Very good	Very good	Very good	Very good
Locally dense	Poor	Poor	Moderate	Moderate	Good	Moderate	Very good
Ability to localise	Poor	Very poor	Very good	Poor	Very good	Moderate	Very good
Polymorphic	Very good	Very good	Poor	Moderate	Very good	Poor	Poor
co-dominant	Yes	No	Yes	No	Yes	No[1]	Yes
Mutation rate	Rapid	Rapid	Reasonable	Reasonable	Rapid	Reasonable	Reasonable
Throughput	Very low	Very low	Moderate	Moderate	Moderate	Moderate	Very high[2]
Technical costs	Very high	Very high	Very high	Very high	Moderate	Moderate	Very low
Repeatable	Poor	Poor	Good	Very poor	Good	Good	Very good

[1] Inferences on genotypes for AFLPs can be improved using densitometry.
[2] As an example, DNA chips now offer 50,000 SNPs in a single reaction.

Current trends in marker choice: SNPs are becoming the marker of choice. Their disadvantage of being pre-dominantly bi-allelic, therefore with lower information content, is being overcome by the number and density of markers available coupled to their high throughput and low cost. Microsatellites retain a use since they can provide considerable information within a few genotypes (e.g. as required for pedigree assignment) and in some applications this will offset the relative disadvantage of throughput in comparison to SNPs.

is provided by the pedigree (Box 3.4). These options are simply doing what we could do before, but removing some of the limitations.

However, the availability of DNA allows us to measure diversity in different ways, since we can obtain the nucleotide sequence of individuals in specific areas of the genome and identify the alleles that are segregating in a population at each position and the genotypes of each individual. Options for addressing diversity with this information include the following:

Box 3.4. Pedigree expectations and DNA estimation of shared alleles.

When we examine genetic variation using the pedigree we are using expectations, but the use of the DNA genotypes helps to estimate the true situation. Consider two full-sibs, each will receive half their autosomal DNA from their sire, and half from their dam, but they will not receive the same half as their sib, and which allele is passed to each offspring is an entirely random process; therefore on average the two sibs may be expected to have only half of the genes passed by the sire and half the genes passed by the dam in common, and it is this average that is assumed when using the pedigree. In reality this proportion of genes shared may be much larger or much less. This proportion can be estimated directly when markers are available that are distributed throughout the genomes of the parents, and are informative for each parent, so that they are capable of distinguishing which of the sire's two homologous alleles were passed to the offspring and which of the dam's.

More dense and more informative markers make the estimate of true proportions of shared alleles more precise. However a caution: simple estimates of proportion using low density markers may introduce substantial distortion outweighing any potential benefit, and the available observed pedigree remains of primary importance unless very dense informative markers are used. For example, Toro *et al.* (2002) compared inbreeding of 62 Iberian pigs from two related strains either calculated from a pedigree going back 20 generations or with molecular coancestries estimated from 49 microsatellites. The correlation was negative for Guadyerbas (–0.32) and low for Torbiscal (0.19) but substantial for all animals together (0.69). Furthermore, the attempt to infer coancestries from molecular markers gave results severely biased because the inference requires information on the true allelic frequencies of markers in the true base population – and these are usually not known. Slate *et al.* (2004) examined 590 sheep of the Coopworth breed with known pedigree for seven generations and genotyped 101 microsatellites: again the correlation was remarkably low (0.17) concluding that, for the correlation between the genealogical and the molecular inbreeding to be substantial, a considerable number of loci and, more important, a high variance of the genealogical inbreeding values is required. This does *not* demonstrate a lack of relationship but demonstrates the noise attached to molecular estimates, for example Daetwyler *et al.* (2006) shows that in Canadian Holstein the observed log heterozygosity based on 10,000 SNP markers has the expected regression of 1 on log (1-F) where F was calculated from pedigree. In conclusion it should be preferable to use pedigree information whenever available, and limiting the use of markers to verify, correct, complete or even implement pedigree recording (Fernández *et al.*, 2005).

1. Examining the diversity in allele frequency, by defining an allele frequency for an individual as 0, ½ or 1 depending on whether it carries 0, 1 or 2 copies of the allele. This trait can be treated as it was a continuous trait, and diversity measured, both between and within breeds, as described above. The idea of individual allele frequency as a trait is an important one, and the trait has a useful property that all

genetic variation is additive i.e. $\sigma_A^2 = \sigma_G^2$. Note that in this approach the breed mean is the estimate of the allele frequency for the breed. As an example, if two breeds are fixed for different alleles then no diversity will be observed within breeds and all the diversity will lay between breeds. This is expanded upon below in paragraph 6 and in chapter 4.

2. The breed means for the frequencies of several alleles, usually from unlinked loci, are combined by some pre-defined function to measure what is called a genetic distance between the breeds. There are several such distance measures (chapter 5) and they often differ in principle from (1) because no explicit consideration is made of variance within breeds. This will be expanded upon in chapter 5.

3. Instead of using gene frequency, the frequency of heterozygotes may be measured. A heterozygote has two different alleles at a locus, and is function of the allele frequencies and non-randomness of mating and survival rates. The justification for this is that in the absence of diversity there will be no heterozygotes in the population. Providing mating is at random, the heterozygosity increases with the number of alleles found in the population and with decreasing variation among the allele frequencies for the population. The assumption of random mating is important and is often assumed to hold within a breed but deviations can be significant, particularly if there is relatively little exchange among breeders. If genotype data is available, the assumption of random mating can be tested by looking at the magnitude and significance of departures from Hardy-Weinberg equilibrium (Falconer and Mackay, 1996; Lynch and Walsh, 1998). Nevertheless in the absence of random mating, both the observed and expected heterozygosity can be informative for studying diversity.

   Whatever result is calculated for the heterozygosity it will depend on the sample of loci used, and extension to inferences about the entire genome are difficult. For example, Toro *et al.* (2006) describes a remark attributed to Nei, that a first approximation to the correlation between the heterozygosity of a sample of r loci and a genome of n loci is $(r/n)^{1/2}$ i.e. for a sample of 20 loci from 20,000 the correlation might be expected to be 0.03. Therefore reliable comparisons between breeds must be made on very dense sets of markers, which may be possible in some species, such as cattle where DNA chips can contain in excess of 50,000 markers.

4. A further simple but limited measure of diversity is counting the number of different alleles appearing in the population for a set of loci, with the more alleles the more diverse. Counting the number of alleles in each breed and the number shared with each other breed offers an opportunity of examining differences between breeds. A variation on this is to count the number of 'private alleles', where a 'private allele'

is defined as an allele found in one breed but in no other. However the heuristic diversity between breeds will be determined not only by whether or not alleles are shared between breeds but also by whether those shared are at similar frequencies within each breed. Therefore the counting approach to measuring diversity appears less valuable than measuring the allele frequencies themselves. Nevertheless observations on private alleles can be very useful in other ways, for example in traceability schemes.

The items (1) to (4) above give some simple ideas about how molecular diversity might be measured. It is a reasonable question to ask what the relationship between molecular and quantitative measures of variation might be, and whether these tell the same or different stories. The answer appears to suggest that the stories are not the same: in a meta-analysis Reed and Frankham (2001) suggest that the mean correlation between molecular and quantitative estimates of diversity is weak (0.22 ± 0.05), indicating that molecular measures of diversity only explain 4% of the variation in quantitative traits. This estimate will include studies that pre-date much of the explosion of molecular data, with the design shortcomings that follow from this limitation. Therefore the question remains open and it is very possible that as dense genome information becomes more available over the next decade, and we gain experience in interpreting it, then our predictions may (should!) improve.

5. Genome-wide patterns of diversity

In paragraph 4 the ways in which information on DNA sequences expanded our views of diversity were explored. However whilst the approaches described above could be applied to any set of markers, the interpretation of the outcome will depend on the positioning of the DNA markers used. The 'functional' DNA in the genome comprises less than 5% of the total DNA (e.g. Federova and Federov, 2005). This includes coding regions for amino acids to be used in proteins, and promoter regions, which control the transcription of the coding DNA. The remaining anonymous DNA includes regions that science has yet to find a purpose for, and may include DNA that is truly without function. For some loci in functional regions an allele may confer a selective advantage on carrier individuals so that the allele will most likely increase in frequency over time and become fixed in the population. It is typically assumed in publications that markers using anonymous DNA are neutral i.e. are not associated with alleles that confer significant selective advantage. The issue over the neutrality of the markers is important since it is assumed that these markers change in frequency only by genetic drift, rather than by drift *and* selection. The neutrality of a locus may differ between breeds since: (1) one breed may have important alleles segregating that are not segregating in another;

and (2) different livestock breeds will be subject to different selection criteria, with these largely dominated by the selection objectives of the breeders concerned.

The genome is organised into chromosomes and this introduces the phenomenon of linkage (Box 3.5). One consequence of linkage is that alleles that are on the same chromosome and close to a new favourable mutation will be tend to increase in frequency alongside the mutation in a process termed 'hitch-hiking' (Maynard-Smith and Haigh, 1974). It is very likely that the alleles very closely linked to the mutation will also become fixed in the population. Therefore this region of the chromosome, very close to the locus under selection, will display very low diversity in the neighbouring loci within a breed. An examination of allelic diversity throughout the genome may

Box 3.5. Chromosomes and linkage.

During fertilisation an individual receives two copies of DNA at each locus, ignoring sex-linked loci, with one copy passed in the gamete received from each parent. The loci are passed in discrete blocks of DNA called chromosomes, and the chromosome passed from the sire and the corresponding chromosome passed from the dam are termed a homologous chromosome pair. The number of human chromosome pairs, including the sex-linked pair, is 23, and the corresponding number in cattle is 30. The chromosomes passed by the parent will be a random choice among the pair of homologous chromosomes it carries, either its own paternally-inherited or maternally-inherited chromosome. The choice is made during the process of gamete formation, called meiosis. The process of meiosis involves crossovers, in which segments of the two homologous chromosomes carried by the parent are exchanged. As a result, new gene sequences may arise. However there are relatively few crossover events relative to the total number of genes and so paternally (or maternally) inherited alleles at neighbouring loci on a chromosome are more likely to be passed to offspring together on the same gamete. However the further away the loci are on the same chromosome, the closer the probability of these alleles being passed together resembles ½ i.e. the probability appropriate for loci on different (heterologous) chromosomes, since there is an increasing chance that an odd number of crossovers has occurred between them. This tendency for paternally (or maternally) inherited alleles on the same chromosome to be passed together is called 'linkage', and two loci on the same homologous chromosomes are said to be linked. Linkage is measured in Morgans (M), or centi-Morgans (cM), and a chromosome's length is defined by the linkage of two loci at either end of the chromosome. A total chromosome length x M is expected to have x crossovers during meiosis, although the number that actually occurs is a random variable (see Lynch and Walsh, 1998 for further information on linkage and measuring linkage). A typical mammalian chromosome has a length of 1 M. Two loci a distance of 0 M apart indicates that the same parent's alleles are invariably passed together, and an infinite distance apart indicates a probability of ½ of being passed together.

therefore show patterns of regions of high diversity punctuated by regions of relatively low diversity. This pattern of diversity *within* the genome is called a selection footprint and may indicate loci important for domestication, or for the characteristics of particular breeds (e.g. Wiener *et al.*, 2003), or simply highly-conserved regions for the genus as a whole, whether wild or domesticated. These regions will be further discussed in more detail in chapter 4. Effective searching for selection footprints is only just beginning in livestock species with the availability of dense, affordable, genome-wide markers such as SNPs.

More generally, the expansion in DNA information will allow the diversity of allelic combinations at loci distributed throughout the genome to be studied. This type of diversity within breeds will depend not only on the allele frequencies but also on the extent of linkage disequilibrium (LD) that is observed. This LD may arise from the breed history of census size and management over time, including bottlenecks or introgression. However this form of diversity, both between and within breeds, may indicate the presence of epistatic interactions affecting performance. In summary, the study of genetic diversity will extend to differences in genomic patterns between and within breeds.

6. Measuring changes in diversity

So far we have simply considered measuring the amount of diversity present in a population. This will reflect events in the history of the population or the breed, often long ago, and the information on the diversity may illuminate the breed origins. However, sustainable management of genetic resources is concerned with managing the diversity that is present today. It is important to realise that some genetic variation is inevitably lost in each generation due to the inherent randomness in the passing of alleles from parent to offspring. It is impossible to ensure that every distinct variation at each of ~ 3×10^9 base pairs of a genome can be replicated in the individuals selected as replacements for the current generation. Nevertheless in each generation there is potentially new variation entering the population as a result of mutation, immigration (if the population is not closed) or the influence of epistatic interactions uncovering new variations (Carlborg *et al.*, 2006) in selected populations. Therefore the sustainable management is more concerned with maintaining the expected rate of loss of existing variation to a sustainable level, justified in more detail by Woolliams *et al.* (2002). Therefore what is required is a means of measuring or predicting the rate of loss.

The important concept in measuring the rate of loss is the idea of inbreeding. To measure inbreeding we identify a reference point in the history of a population, called the base generation, when we assume that all the alleles at an assumed neutral locus in this generation are identifiably different. This is of course unrealistic, but these assumptions provide the necessary conceptual framework to predict the real changes in diversity that we observe with time! Every individual born after the base generation will then have an 'inbreeding coefficient' determined by its pedigree, as described in Box 3.6, and these coefficients will directly determine the expected loss of diversity relative to the base generation.

The following points are important for considering inbreeding. For a single population derived from a base generation at t=0, and where $\sigma_A^2(t)$, H(t) and F(t) denote the values of σ_A^2, heterozygosity at time t, and mean inbreeding coefficient at time t:

1. The rate of inbreeding, $\Delta F = [F(t) - F(t-1)]/[1 - F(t-1)]$ is a multiplicative definition, and is constant for a population of constant size and subject to a constant selection regime. Note $F(t)-F(t-1)$ is not constant in this case.
2. Equivalently $\Delta F = [H(t-1)-H(t)]/H(t-1)$, i.e. the fractional loss of heterozygosity in a generation.
3. It is expected that $H(t) = [1-F(t)]\, H(0)$ i.e. heterozygosity will decrease if averaged over many lines inbred from the same population. However the process is a random one and heterozygosity may increase or decrease if observed in only a single line.
4. For a trait where $\sigma_A^2(0) = \sigma_G^2(0)$, it is expected that $\sigma_A^2(t) = [1-F(t)]\, \sigma_A^2(0)$. However as with item (3) it is an expectation not a rule.
5. Where $\sigma_A^2(0) = \sigma_G^2(0)$, the genetic variance between isolated sub-populations drawn from the same base generation is given by $2F(t)\sigma_A^2(0)$. Since an allele's frequency is a trait where $\sigma_A^2(0) = \sigma_G^2(0)$, this observation shows that the frequency of a neutral allele will drift away from the initial frequency with accumulated variance $2F(t)\sigma_A^2(0)$, and this variance is called the *drift variance*. This drift over time of allele frequencies makes it feasible that alleles will either be lost or fixed in a population; in fact over a sufficiently long and indefinite period, it is certain that an allele will either become lost or fixed.

Items (1) to (5) indicate that changes in some important measures of diversity per unit of time are described by ΔF.

It will be explained in chapter 7 that many aspects of sustainable management within breeds will depend on managing ΔF, and so this parameter is very important. Chapter 7 will also explore some predictive formulae appropriate for different conditions. ΔF is often reported as a transformed value called the 'effective population size', $Ne = [2\Delta F]^{-1}$, where ΔF is the rate of inbreeding measured over 1 generation of the population. In general, for most livestock populations the number of parents will greatly exceed the calculated value of Ne, however the justification for its use is that Ne diploid (single-sex) individuals would have an identical ΔF if they were subject to random selection and

Box 3.6. Inbreeding coefficients and examples

The inbreeding coefficient is defined with reference to a base generation in which all the individuals are assumed unrelated and that all the alleles in the base generation are considered distinct. Thus for a base generation with a total of N parents, there are 2N distinct alleles, since each parent carries 2 alleles. The inbreeding coefficient, F, of an individual is then defined as the probability that for a randomly-chosen neutral locus the two alleles carried by the individual are identical by descent, i.e. copies of the same allele from the base population.

There are a number of properties of inbreeding coefficients arising from this definition:

1. F is a probability and so $0 \leq F \leq 1$.
2. F = 0 for individuals in the base generation, by definition
3. For species with no selfing, F > 0 only when there is a common ancestor in the pedigree of the sire and the pedigree of the dam; equivalently, F > 0 when there is a loop in the genealogical tree of the individual's pedigree.

As an example of calculation consider the pedigree below.

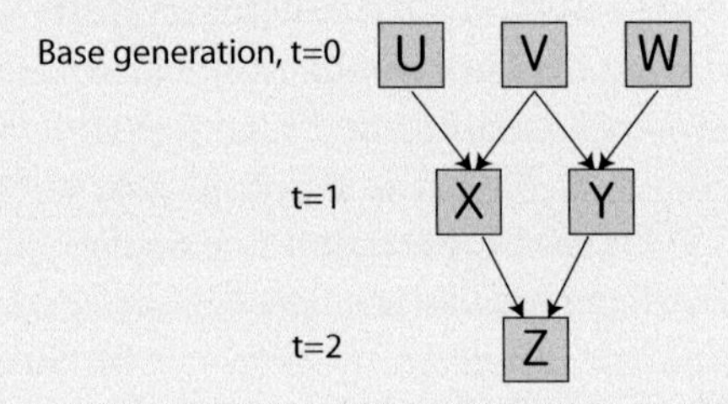

F = 0 for 'U', 'V', 'W', 'X', 'Y'. For 'Z', 'V' is an ancestor common to both sire and dam so F > 0, with a loop defined by 'V' → 'X' → 'Z' → 'Y' → 'V'. If 'Z' has two copies of the same allele from the base, then three events must all have occurred: (a) 'V' must have passed the same allele to both 'X' and 'Y', which occurs with probability ½; (b) this allele must have been passed to 'Z' by 'X', which occurs with probability ½; and (c) this allele must have been passed to 'Z' by 'Y', which occurs with probability ½. Therefore the probability that 'Z' has two alleles that are identical by descent is $\frac{1}{2} \times \frac{1}{2} \times \frac{1}{2} = 1/8$.

The methodology for calculating inbreeding coefficients in complex pedigrees will not be given here and readers are referred to Falconer and Mackay (1996) and literature describing the calculation of the numerator relationship matrix. Inbreeding coefficients steadily increase in a closed population over time, towards 1.

random mating (including selfing) with no limitations on family size. This analogous population is often called the 'idealised population', but it is very misleading to consider other genetic properties of the idealised population as being analogous to the real one! Therefore effective population size is a useful device for visualising what a rate of inbreeding may mean in an approachable way. Beware, do not accept any other definition for Ne other than Ne = $[2\Delta F]^{-1}$, even if stated in a scientific publication or text book – they are wrong and potentially very misleading! See Box 3.7 for a description of issues surrounding the use of Ne.

7. Relating diversity measures to F and ΔF

If we re-examine the measures of diversity described in paragraph 4, we may use the inbreeding framework of paragraph 6 to inform our inferences in a number of ways:

1. Heterozygosity at an initial point will be completely determined by the allele frequencies at the chosen loci and prior breed history, but the change in heterozygosity is related to ΔF and the number of generations over which the change is measured. The observed heterozygosity and change in heterozygosity as measured by markers is

Box 3.7. Pitfalls in effective population sizes.

1. A common unit of time in considering livestock populations is the year, since for many species and for many farmers it constitutes the length of the husbandry or economic cycle. However the populations are renewed at different rates called the generation interval (Falconer and Mackay, 1996), and for common livestock species this may be much longer e.g. ~3 years for sheep, ~6 years for cattle, ~10 years for horses. Data over time will often appear in units of years, and consequently when estimating ΔF the parameter will then appear as scaled per year, say ΔF(y). It would be tempting, but dangerous to calculate $Ne(y) = 1/[2\Delta F(y)]$. It is more appropriate to calculate $Ne(g) = 1/[2\Delta F(g)]$ where 2ΔF(g) is the rate of inbreeding per generation, since this more truly reflects the expected loss of diversity incurred in replacing a generation. For a generation of L years $\Delta F(g) \approx L\,\Delta F(y)$, and $Ne(g) = 1/[2L\Delta F(y)]$. The impact of this can be seen by considering a horse population where L = 10 years and ΔF(y) = 0.0025, and a pig population where L = 1 year and ΔF(y) = 0.01. For these, Ne(y) = 200 and 50 for horse and pigs respectively, whereas Ne(g) = 20 and 50 respectively. Comparison of Ne(y) suggests the population at greater genetic risk is the pigs, yet comparison of Ne(g) makes it clear that the population subject to greatest genetic risk is the horse population. See chapters 7 and 8 for further discussion of generation interval.
2. It is common and *erroneous* to give an unqualified definition of $Ne = 4MF/[M+F]$ where M and F are the numbers of male and female parents. From the correct definition of Ne this is equivalent to stating $\Delta F = [8M]^{-1} + [8F]^{-1}$, a predictive formula derived by Wright (1969) that is *only* appropriate where there is random selection and mating among the 2 sexes with unrestricted family sizes. Such a situation rarely applies and very frequently such an estimate will be a gross underestimate of ΔF.

subject to considerable sampling error and, consequently, the change in inbreeding coefficient may be low. Hence changes in heterozygosity are not particularly robust for measuring effective population size, although studies such as Daetwyler *et al.* (2006) show that observed heterozygosity in field populations has the expected relationship with F(t) as stated in paragraph 6.3.

2. The diversity in allelic frequency between and within breeds can be related to inbreeding providing we assume that at some time in the past the group of breeds that are being studied are all derived from the same base population, and that all the alleles were present in the base population, and are assumed to be neutral in all populations involved. This second assumption is strong. An allele may be present in one group of breeds but absent in another group because (1) the allele was present in the base but has been lost in some breeds (see paragraph 6 item5). (2) an allele appeared as a mutation in a breed that contributed to the development of one group of breeds but not to the other group, or (3) a mutation occurred several times in different breeds. The likelihood of these different options will depend on how far back to the common base, and what type of marker, since some marker types, such as microsatellites (Box 3.3), may be subject to faster mutation rates than others making it more likely that some alleles will have appeared in recent times. In livestock, hitch-hiking associated with selection (Maynard-Smith and Haigh, 1974) may raise questions on the extent of neutrality over anonymous DNA markers over long periods of time.

   However, if these assumptions are made, we can relate the observed σ_B^2 for allele frequencies to the value of $2F(t)\sigma_A^2(0)$ (see paragraph 6 item5), where $\sigma_A^2(0)$ is an estimated genetic variance for allele frequency in the presumed base generation, and F(t) is the estimated inbreeding of the populations relative to a base from which the breeds are assumed to be distinct sub-lines. This concept leads after further development to the measure of breed relationships called F_{ST} developed in chapter 5, which decomposes the estimated inbreeding of two individuals from two breeds into that part of the inbreeding process shared by two breeds, and the remainders.

3. In Box 3.2 the concept of variation between breeds in performance traits was explored. This estimate of breed variation is based upon the observed variation present now and tells us how much performance may be improved or reduced from breed substitution. However McKay and Latta (2002) review Q_{ST}, a measure of breed relationship developed by treating the observed genetic variation between and within breeds for quantitative traits as if it were derived by drift from a base population, i.e. analogous to the development of F_{ST} for the variation in allele frequency. The

interpretation of this measure in terms of evolutionary forces is difficult in livestock because selection plays a major role in breed development and neutrality of easily measured quantitative traits (conformation, production, reproduction) is perhaps more questionable than anonymous DNA markers. Q_{ST} can be expressed as $\sigma_B^2/(\sigma_B^2+2\sigma_W^2)$, so high values of Q_{ST} indicate a high degree of differentiation between breeds, and consequently the importance of breed variation. It is closely related to the direct measure $g_1=\sigma_B^2/(\sigma_B^2+\sigma_W^2)$ based upon extant genetic variation since $Q_{ST} = g_1/(2-g_1)$. For example, using the values of g_1 in Box 3.2, estimates of Q_{ST} for feed conversion and relative growth rate were 0.14 and 0.20.

4. Probabilities of fixation and loss of individual alleles from a population and hence the number of alleles we observe will depend on migration, drift and also selection if the allele is not neutral. The extent of drift, and consequently how these probabilities will change over time, will depend on ΔF. This is covered in more detail by Crow and Kimura (1972), and results have also been developed for selection (e.g. Caballero *et al.*, 1996).

8. Conclusion

In 'The Name of the Rose' by Umberto Eco (1992, as translated by William Weaver), Father William of Baskerville delights in the diversity of nature by declaring 'the beauty of the cosmos derives not only from unity in variety, but also from variety in unity'. He summarises in just 18 words – and only 13 distinct words – that when examining the population of a species we should expect variety, and will invariably find it, and as we look closer at what appears at first sight to be a more uniform sub-population of individuals, so variety remains. However, quantifying diversity, understanding its scientific nature and importance, and providing guidance on how to use and conserve it, requires many more words!

References

Caballero, A., M. Wei and W.G. Hill, 1996 Survival rates of mutant genes under artificial selection using individual and family information. Journal of Genetics 75: 63-80.

Crow, J.F. and M. Kimura,1972. Introduction to Population Genetics Theory. Harper & Row.

Carlborg, O., L. Jacobsson, P. Ahgren, P. Siegel and L. Andersson., 2006. Epistasis and the release of genetic variation during long-term selection. Nature Genetics 38: 418-420.

Cundiff, L.V., M.D. MacNeil, K.E. Gregory and R.M. Koch, 1986. Between- and within-breed genetic analysis of calving traits, and survival to weaning in beef cattle. Journal of Animal Science 63: 27-33.

Daetwyler, H.D., F.S. Schenkel and J.A.B. Robinson, 2006. Relationship of multilocus homozygosity and inbreeding in Canadian Holstein sires. Proceedings of the Canadian Society of Animal Scientists Halifax Joint Colloquium, Halifax, NS, Canada.

Eco, U. and W. Weaver, 1992. The Name of the Rose. Vintage.

Falconer, D.S. and T.F.C Mackay, 1996. Introduction to Quantitative Traits. 4th edition, Longman, Harlow, Essex, UK.

Fedorova L. and A. Fedorov, 2005. Puzzles of the human genome: Why do we need our introns? Current Genomics 6: 589-595.

Fernandez, J., Villanueva, R. Pong-Wong and M.A. Toro, 2005. Efficiency of the use of pedigree and molecular marker information in conservation programs. Genetics 170: 1313-1321.

Jenkins, T.G., M. Kaps. L.V. Cundiff and C.L. Ferrell, 1991. Evaluation of between- and within- breed variation in measures of weight to age-relationships. Journal of Animal Science 69: 3118-3128.

Lush, J.L., 1994. The Genetics of Populations. Special Report 94. Iowa State University.

Lynch, M. and B. Walsh, 1998. Genetics and Analysis of Quantitative Traits. Sinauer Associates.

McKay, J.K. and R.G. Latta, 2002. Adaptive population divergence: markers, QTL and traits. Trends in Ecology and Evolution 17: 285-291.

Maynard Smith, J. and J. Haigh, 1974.The hitch-hiking effect of a favourable gene. Genetical Research 23: 23–35.

Reed D.H. and R. Frankham, 2001. How closely correlated are molecular and quantitative measures of genetic variation? A meta-analysis. Evolution 55: 1095-1103.

Slate, J., P. David, K.G. Dodds, B.A. Veenvliet, B.C. Glass, T.E. Broad and J.C. McEwan, 2004. Understanding the relationship between the inbreeding coefficient and multilocus heterozygosity: theoretical expectations and empirical data. Heredity 93: 255-265.

Thiessen, R.B., E. Hnizdo, D.A.G. Maxwell, D. Gibson and St.C.S. Taylor, 1984. Multibreed comparisons of British cattle: variation in body-weight, growth-rate and food-intake. Animal Production 38: 323-340.

Thiessen, R.B., St.C.S. Taylor and J. Murray, 1985. Multibreed comparisons of British cattle: variation in relative growth-rate, relative food-intake and food conversion efficiency. Animal Production 41: 193-199.

Toro, M., C. Barragan, C. Ovilo, J. Rodriganez, C. Rodriguez and L. Silió, 2002. Estimation of coancestry in Iberian pigs using molecular markers. Conservation Genetics 3: 309-320.

Toro, M., J. Fernandez and A. Caballero, 2006. Scientific basis for policies in conservation of farm animal genetic resources. Proceedings 8th World Congress on Genetics Applied to Livestock Production, CD-ROM Communication No. 33-05.

Wheeler, T.L., L.V. Cundiff, R.M. Koch and J.D. Crouse, 1996. Characterisation of biological types of cattle (cycle IV): carcass traits and longissimus palatability. Journal of Animal Science 74: 1023-1035.

Wiener, P., D. Burton, P. Ajmone-Marsan, S. Dunner, G. Mommens, I.J. Nijman, C. Rodellar, A. Valentini and J.L. Williams., 2003. Signatures of selection? Patterns of microsatellite diversity on a chromosome containing a selected locus. Heredity 90: 350-358.

Woolliams, J.A., R. Pong-Wong and B. Villanueva, 2002. Strategic optimisation of short and long-term gain and inbreeding in MAS and non-MAS schemes. Proceedings of the 7th World Congress on Genetics Applied to Livestock Production 33: 155-162.

Wright, S., 1969. Evolution and the genetics of populations. Vol. 2. The Theory of Gene Frequencies. University of Chicago, Chicago.

Chapter 4. Genomics reveals domestication history and facilitates breed development

Miguel Toro[1] and Asko Mäki-Tanila[2]
[1]Department of Animal Breeding, National Agriculture and Food Research Institute, Carretera La Coruna km7, 28040 Madrid, Spain
[2]Biotechnology and Food Research, MTT Agrifood Research Finland, Jokioinen, Finland

Questions that will be answered in this chapter:

- *What is known about animal domestication sites and time points?*
- *How could the historical breed events be revealed by genomic studies?*
- *What is the value of this knowledge for utilisation and conservation?*
- *How are breed differences utilised in improving animal production?*
- *How can the effectiveness of breeding and conservation be enhanced with molecular genetic tools?*

Summary

The first part of the chapter deals with domestication and describes what is known from archaeological and historical records about the main sites and time points for the domestication of different species. This includes a review of why some species are suitable for human use and amenable to domestication and some are not. Then, more recent developments are examined which led to the formation of the breed structure we observe today. It includes a review of the main principles in breed formation, and how some breeds have become the predominant, through intense selection combined with the use of modern technology. The second part introduces genomic methods that are being used to reveal how selection, population demography and admixtures have intervened in the development of farm animal populations. The last section in this chapter is devoted to the implementation of the knowledge on breed differences with emphasis on using genomic tools for detection and utilisation of QTL information and for assigning individuals and products to breeds.

1. Animal domestication

1.1. Domestication

Domestication of plants and animals has been one of the most important events in the history of mankind. It has increased the amount of available food and, as consequence, has supported the growth of human populations and their capability to expand into new environments, and has heavily influenced mankind's cultural evolution. From a genetic point of view, animal domestication has involved selection for reduced aggressiveness, earlier sexual maturity, tolerance for living in confinement, and a number of morphological traits. As an example of the latter, chickens were probably selected for increased size, whereas cattle for a smaller size. In this process there were significant changes in both the traits selected and in correlated traits. The most notable ones were the decreased brain size and the size reduction of teeth (Diamond, 2002).

Altogether some twenty terrestrial and a few fish species have become adapted to being bred and fed in captivity, and to satisfying the diverse human needs. There are several factors that are genetically differentiating animal populations:

- separate founder populations within species;
- isolation accompanied by variation in the number of breeding animals;
- human controlled environments, mostly relaxing natural selection;
- harsher environments in new regions, often strengthening natural selection;
- human selection in managed populations targeted towards mankind's needs and objectives.

Domestication is usually defined as a process in which populations adapt to mankind and its environment. Yet it may be also considered as a form of mutualism involving a parallel evolution in cultures and genomes. An example of how important this co-evolution can be, is the domestication of milk-producing cattle and the use of milk. This has resulted in selection of both animals and humans (Beja-Pereira *et al.*, 2003). In mammals, milk usually has little nutritional value for adults as the lactase enzyme, necessary for digesting lactose in milk into galactose and glucose, is turned off at weaning. However, about 30% of mankind is now lactose-tolerant and this condition is common in the northern latitudes in Europe where milk provides a rich source of calcium, vitamin D and protein. Lactose-intolerance was the normal ancestral condition but a 'recent' mutation in the lactase gene, maintaining the ability to digest lactose through adulthood, was quickly favoured by natural selection in human populations that raised dairy cattle.

1.1.1. Why were some species domesticated and others not?

Animal domestication started with dogs, sheep and goats, and continues in recent times with fur animals and fish species such as the salmon. There are 8-10 centres of origin for plant and animal domestication and food production (Table 4.1). Authors from Galton (1883) to Diamond (2002) have questioned why only a dozen of species among the 148 species of large terrestrial mammalian herbivores and omnivores, all plausible candidates for domestication, have been domesticated. Diamond (2002) claims that the obstacles are not human abilities but characteristics of the target species itself. He quotes six main obstacles (with an example of a species between brackets):

- a specialised diet not easily supplied by humans (anteater);
- a slow growth rate and a long generation interval (elephant);
- a nasty disposition (grizzly bear);
- the reluctance to a breed in captivity (cheetah);
- a lack of follow-the-leader dominance hierarchies (antelope);
- a tendency for extreme panic in enclosures when facing predators (gazelle).

Table 4.1. Domestication events, time points and sites for farmed animal species, based on the reviews by Bruford *et al.* (2003), Mignon-Grasteau *et al.* (2005), Dobney and Larson (2006) and Zeder *et al.* (2006). There are discrepancies between the estimates based on archaeological and molecular (not presented) evidence.

species	no. events	domestication site	archaeological dating (years BC)
dog	many	East-Asia	15,000
sheep	1-3	Near East	12,000
goat	4	Near East	8,000-10,000
pig	7	Near East, Far East, Eurasia	9,000
cattle, zebu	2-3	Near East, India, Africa	2,000-8,000
chicken	1?	Central Asia	5,000-7,000
horse	many?	Eastern Europe, Central Asia	6,000
donkey	2	North Africa	5,000
water buffalo	1-2	South-East Asia	6,000
llama, alpaca	2-4	South America	6,000
camel	?	Near East	3,000
rabbit	?	Europe	2,000

For example, the European horse breeders have tried over several centuries to domesticate zebras after settling in South Africa in the 17th century. Finally, they abandoned the idea, because zebras have an incurably vicious habit of biting the handler and not letting go (they injure more zoo-keepers that tigers), and they have a better peripheral vision than horses making the use of lassoes impossible.

1.1.2. Are there genes for domestication?

Darwin (1868) was the first to note that most domesticated species have undergone similar changes during domestication: appearance of dwarf and giant varieties, piebald coat colour, depigmentation, curly hair, rolled tail, floppy ears and early sexual maturity. Many, but not all, characteristics associated with domestication seem to be linked to pedomorphosis: the retention of juvenile characteristics in the adult body. This suggests that domestication may be the result of changes in a relatively small number of regulatory genes affecting development.

There has been searches for genes specifically responsible for these traits, e.g. for genes underlying differences in coat colour pattern in pigs and horses, plumage in chickens or muscle mass in pigs (Andersson and Georges, 2004). In the 1950s, Belyaev (see Trut, 1999) started a selection trial for tameness in the Silver Fox. He hypothesised that morphological, physiological and behavioural traits were simultaneously modified by domestication and that selection for an important behavioural trait would modify the others. Belyaev measured tameness by the ability of young sexually mature foxes to behave in a friendly manner towards their handlers, by wagging their tails and whining. After more than forty years of selection, 70-80% of the test population accepted human contact and often licked the persons looking after them, like dogs do. As predicted by Belyaev, additional changes appear such as piebald coat colour, drooping ears and shorter tails and snouts. Physiological changes also occurred: in domesticated animals the corticosteroid levels rise significantly later and they have much lower adrenal responses to stress and more serotonin in their blood.

1.1.3. Multiple versus single domestication

It has been a long debate, whether domestic animals are the result of a single domestication event in a restricted geographical area or of multiple, independent domestication events in different geographical regions. The ample distribution of domesticated Eurasian mammals from Portugal to China supports the idea of different independent domestications. This has been confirmed by molecular genetic data (see reviews: Bruford *et al.*, 2003, Mignon-Grasteau *et al.*, 2005, Dobney and Larson, 2006, Zeder *et al.*, 2006), but the answer depends on species. In some cases (pigs and horses) the hypothesis of more than two domestication events is well supported by the scientific data, whereas in the case of cattle (Loftus *et al.*, 1994) and donkeys the evidence points to two (Table 4.1).

Tracing mitochondrial DNA (mtDNA, Box 4.1) is the standard method for tracking the history of domestication. It is well known that cattle can be classified into two groups: zebu (*Bos indicus*) from East Eurasia and East Africa, and taurine (*Bos taurus*) from Europe and Mid, North and West Africa. Looking at mtDNA analyses, the two breed groups are very distant suggesting two separate domestications from differentiated subspecies of the auroch (*Bos primigenius*) cattle. The African zebu cattle only have taurine mtDNA markers but nuclear and Y chromosomes markers are similar to those of the Asian zebu. In this case mtDNA studies were not able to detect zebu genes that were passed by zebu bulls to a few taurine female founders (Bruford *et al.*, 2003, Zeder *et al.*, 2006). The mtDNA studies have to be supported by analysis of nuclear DNA data to get clear results.

In sheep and goats there is one major geographical mtDNA lineage that probably represents initial domestication in the Fertile Crescent, with two more restricted lineages representing later independent domestication events. In goats, the first line

Box 4.1. Molecular markers for studying domestication.

Mitochondrial DNA polymorphism

Variation in mtDNA is extremely useful in studying genetic diversity. There are a number of reasons for this:

- mtDNA is maternally inherited with no recombination and, all markers in this genome are effectively linked as a single haplotype. Hence the number of nucleotide differences between mitochondrial genomes is a direct reflection of the genetic distance that separates them;
- each cell has thousands of copies of mtDNA; and
- regions of mtDNA mutate 5-10 times more than nuclear DNA making it ideal for studying the divergence between wild and domestic populations under the relative short timescale of domestication (<10,000 years).

The usual way to analyse mtDNA is the sequencing of the cytochrome b gene and of the control region that shows greater variation than the other parts of the mtDNA molecule.

Y chromosome

In the same way as mtDNA could be used to identify maternal lineages in the populations, Y chromosomes sequences provide similar information on paternal lineages. There, recombination is restricted to a small area of the chromosome.

expands from the Fertile Crescent all over the world. Another, limited to West Pakistan, is related to cashmere lines, whilst the last is of uncertain origin (Bruford *et al.*, 2003, for sheep; see also Tapio *et al.*, 2006). In pigs the first mtDNA studies suggest two domestication events (Asia and Europe), but the most recent mtDNA study suggests up to seven events across Eurasia (Larson *et al.*, 2005). Dog history appears to resemble that of pigs rather than of cattle, with multiple events in several locations occurring through domestication of wolves. In contrast, the domestication of horses was neither limited in time nor space. There is extensive matrilinear diversity seen in mtDNA variation that does not match with a patrilinear diversity of the male specific Y chromosome. This is consistent with a strong sex-bias in the domestication process with only a few stallions contributing genetically to the domestic horses (Lindgren *et al.*, 2004). Donkey seems to be the only ungulate domesticated solely in Africa and probably as the result of two domestication events (Beja-Pereira *et al.*, 2004).

The South American camelids present a complex situation. The modern-day alpaca and llama could descend either from the wild guanaco or from the wild vicuña. Both llama and alpaca have mtDNA haplotypes from both these wild species whilst microsatellites indicate a close relationship between alpaca and vicuña and between llama and guanaco, respectively. This implies differences in how mtDNA and microsatellite variation are revealing the history, and in the future data on the Y chromosome will help to clarify the picture (Zeder *et al.*, 2006).

1.2. Breeds and modern genetic improvement

Developing farm animal populations into breeds or pure lines is a very recent event from a domestication history perspective. There were numerous local populations that provided the basis for the breeds in the 19th century. The sequence of events started with a relatively high market value of a specific group of animals and continued towards the demand for a purebred pedigree. This demand represents an initial step towards scientific breeding as a means of reducing risks in breeding either for the market or as a hobby. Domestic trade and export of animals have stimulated the efforts to set the formal definition of a breed, although this remains very difficult (chapter 3, Box 3.2). Livestock shows have had an effect on emphasising the type and ideas about the correct conformation and colour. Therefore breeds have been selected for exterior traits (confirmation, colour, horns) with reasons which may have been quite weak and occasionally with quite serious consequences for welfare (e.g. in companion animals). The selection has been accompanied with assortative mating and intentional inbreeding, which can also bring unforeseen and unwelcome consequences. However these risks have been often successfully avoided, for example the famous Shorthorn breed was founded by extensive use of only a very few bulls. There do exist breeds that do not differ genetically much from each other in performance and production. They are being labelled different only because of distinct visible phenotypes, or *vice versa*, phenotypic similarity of breeds due to favouring of a similar colour pattern may mask surprisingly large amounts of underlying genetic diversity.

In dogs, the extraordinary variation in shape, size, behaviour and physiology of the breeds makes them a unique genetic model. Modern dog breeds have been generated by selecting for many particular and diverse traits desired by mankind among the wild ancestors of the dog. Each pure breed is an inbred, isolated genetic population, with simplified genetic structures that can be linked to their physical traits.

In terms of conserving the total genetic diversity, theoretically one of the most promising ways to maintain variation is creation of a large number of inbred lines as long as inbreeding does not compromise their survival (chapter 7). Breeds serve as partially inbred lines and their composition harbours lots of genetic variation with a low risk for the erosion of the total diversity. This is merely a corollary from breed formation, not influenced by the modern thinking about cost-efficient secure ways to maintain diversity.

For animal populations, where the diversity in individuals' outlook has been retained, a term primitive breed has been used. For example, European goat breeds are phenotypically very heterogeneous. In Iceland, a uniform outlook has never been a target in cattle, sheep or horse, whilst variation in colour has been much appreciated (Adalsteinsson, 1981). Therefore the present Icelandic animal breeds exhibit a wide range of colours in a way not usually found elsewhere in Europe. Similar diversity of exterior traits is almost a norm in farm animal populations in developing countries.

Animal breeding was modernised 50-60 years ago brought about by market growth, transport and communication, an improved understanding of genetics, an increase in reproductive rate through development of reproductive technology and computation power. The empirical selection done earlier has been geared up to very efficient selection programmes with clear objectives, vast bodies of information on animals and utilisation of reproduction technology. The selection is based on quantitative analysis of variation of several production, health and fertility traits.

2. How have the events in breed history modified the genetic variation?

2.1. Marker and sequence information

Genomic research has proven to be a powerful approach in revealing:
- the history of animal populations;
- the number and sites for domestication;
- population expansions and contractions;
- the impact of selection;
- the origin and mixing of maternal and paternal lineages.

The most widely used genetic markers in diversity studies are microsatellites and SNPs (single nucleotide polymorphisms), see Box 3.3 in chapter 3. It is now possible with modern DNA chip technology to analyse even up to thousands of loci, including a vast number of putative neutral loci across the genome. The information can be combined from several (linked) sites and used to follow the combinations of alleles from different loci or haplotypes and the haplotype frequencies in the processes of recombination, selection and drift. Now it is also feasible to obtain complete sequence information on chosen areas of the genome. The outcome can be used to estimate the relatedness of sequence variants and trace them back to ancestral sequences.

2.2. Detection of selection

Genetic mutations that increase mutual benefit to domesticated animals and mankind, give these the individual animals carrying them a selective advantage, helping them to have more offspring, and this will be repeated in the offspring that inherit the mutations. These kinds of mutations are very likely to spread through the population over generations, rather than disappear from the population, and are seen in genomes today. Consequently, it should be clear to find the cases where a particular allele of a locus has been so beneficial that it has spread quickly and widely in population(s) and thereby reduced the variation. The consequence of this spread is called a signature of selection: the level of variability will be reduced and the level of linkage disequilibrium and the genetic differentiation between populations will be increased.

Spotlighting signatures of selection will also face obstacles. It is possible that demographic processes produce similar patterns as selection. Therefore the effects of selection and breed history may be hard to untangle. A high frequency of an allele in one population and complete absence in another related older population may be an outcome of different selection pressures. But it can also be an historical accident - not a mark of selection - if the founders of the population just happened to be carriers, even if uncommon, of the allele, or drift in frequencies by isolated lines.

Detection of selection starts from understanding the behaviour of the genome or parts of it under neutral conditions when selection is not present. Loci under selection will often behave differently and therefore reveal outlier patterns in variation. Predictions for neutral loci in very large populations, which have had the same size over generations, can be obtained and also extended to finite (small) populations. These predictions are usually accompanied by case-specific statistical parameters, and tests applicable for single markers, large sets of markers and sequence data. Often user-friendly software packages are attached to different test procedures. Instead of deducing expectations from population genetics theory, genotyping a large number of neutral loci provides a baseline to test for outliers. Demographic processes, such as migrations, population contractions and expansions, and random drift, affect the whole genome, whereas selection leaves its mark on specific important functional regions in the genome. Using the whole genome as a baseline, it is straightforward to find deviating patterns within the genome.

2.3. Basic methods for finding outliers

2.3.1. The Lewontin-Krakauer test

A recent mutation in a population is first relatively rare. In some populations it can quickly become quite common. Such common "young" alleles can be a sign of selection, because new favourable mutations replace other alleles faster than the neutral ones. When a locus shows extraordinary levels of differentiation between populations (measured by Fst, chapter 5) compared with other loci, this may be interpreted as evidence for selection of an allele in one of the populations. A classic test for selection/neutrality by Lewontin and Krakauer (1973) exploits this fact. The test rejects the neutral model for a locus, if the level of genetic differentiation between populations is larger than predicted. This test has been rediscovered, and new versions have been developed for large-scale genomic data (Akey *et al.*, 2002), and have augmented it by statistical sophistication (Beaumont *et al.*, 2002).

2.3.2. Sequence diversity

Several descriptive measures are used to summarise polymorphisms of DNA sequences. Under a neutral model, the expected level of diversity (commonly symbolised θ) can be deduced from the generation of new alleles by mutations and from the elimination of alleles by drift (which is inversely proportional to effective population size), i.e. $\theta = 4$ x effective population size N_e x mutation rate. There are several measures for the molecular genetic diversity: numbers of alleles, numbers of segregating sites or the number of nucleotide positions at which polymorphism is found (S) and the average number of pair wise differences in a set of sequences (π). For comparison, in a human population two randomly chosen individuals differ at ~1 in 1,000 nucleotides (1 SNP per kilo base). The genetic diversity of mankind is low compared to other (older) species. In cattle and sheep the mean nucleotide diversity is 2-2.5 SNPs per kilo base (cf. Meadows *et al.*, 2004), whilst for the chicken the estimate is 4-5.5 (Hillier *et al.*, 2004).

The outliers or departures from the neutral expectation can be assessed using statistical tests. The average number of pair wise differences is estimated as

$$\pi = \Sigma\, x_i\, x_j\, \delta_{ij} \,/\, n$$

where n is the length of the sequence, x_i is the frequency of sequence type i, x_j is the frequency of sequence type j and δ_{ij} is the number of nucleotide differences between the haplotypes i and j. It is directly an estimate of θ (say θ_π) and the estimate deduced from the number of segregating sites (θ_S) is $S / \Sigma\,(1/i)$ where summing is over n-1. Tajima (1989) has constructed a measure to compare the estimates θ_π and θ_S. With no selection the estimates should be indistinguishable and the test statistic $D = \theta_\pi - \theta_S$ should be zero. A prolonged population bottleneck should reduce S and results in a positive D. Purifying selection will reduce heterozygosity, hence negative D values, *vice versa* positive values will be observed under balancing selection. If a population is expanding, many sequence types are seen. But the contribution to heterozygosity will be low and D will be negative.

2.3.3. Linkage disequilibrium

Linkage disequilibrium (LD) describes a situation in which some combinations of alleles of two or more loci (haplotypes) occur less or more frequently than would be expected from the allele frequencies at the loci. In a way, loci in LD are co-segregating. LD is caused by selection, bottlenecks, migration and mutation. The natural state of a new mutation is in LD since it occurs in one animal in the midst of a single sequence in the population. Related to LD, the genome contains regions with reduced haplotype diversity – called *haplotype blocks* (Wall and Pritchard, 2003) – separated by regions of higher diversity. Their identification is suggested to facilitate whole-genome screenings for interesting genes with fewer markers than when the haplotype blocks are ignored (Johnson *et al.*, 2001). The generation of haplotype blocks is not completely understood. Often they are associated with variation of recombination rate in the genome (e.g. Daly *et al.*, 2001), but it is also shown that blocks may stem form uniform recombination rate and drift only (Zhang *et al.*, 2003).

2.4. Selective sweep

When genome screening is done with a large set of markers covering the whole genome, it is possible to detect common alleles surrounded by extensive linkage disequilibrium. For example, selection is causing correlated changes of allele frequencies in adjacent loci, and its effect can be traced by LD analyses. When and wherever selection acts on a mutation, it will also affect the linked sites leaving its mark in the surrounding chromosomal area. This signature, called selective sweep, is seen as a reduced variation at linked sites of loci under selection. A synonym is genetic hitch-hiking (Maynard-Smith and Haigh, 1974): the change in the frequency of an allele due to selection on a closely linked locus with a positive allele. However, such selection could also be negative. Then, the term background selection is used when selection is removing a harmful mutation and eliminating variation at the adjacent chromosome region (Charlesworth *et al.*, 1993). A selective sweep may have a dramatic impact on the level of population subdivision in a particular genomic region if the sweep is active in only few populations.

The strength of the sweep effect depends on the magnitude of selection. On the other hand, the further away a neutral locus is from a positively selected one, the less affected it would be by selection. The selective sweep effects will be stronger in regions of lower recombination; therefore we may expect linkage distances rather than physical distances to be more relevant. LD decays over many generations of recombination, so only recent selective sweeps can be detected. After a beneficial allele is fixed or an harmful allele is removed, the selective sweep will decay with time due to recombination and therefore selection in the remote past might not be detectable.

2.5. Expansions and contractions

The control region in mtDNA shows extraordinary amounts of variation within species and therefore it is used to track patterns and development of diversity. Its high rate of change and the ability to detect differentiation between domestic breeds makes the control region the method of choice for phylogenetic studies in farm animals. An approach is to examine the distribution of pair wise differences between mtDNA haplotypes within populations (Rogers and Harpending, 1992). If a population has expanded recently, most haplotypes would be separated by only a few substitutions, because there has been no time to accumulate a large number of substitutions with the rate the populations has grown. Also the time point for expansion can be included in the analysis as the populations which have started expanding first, would have the highest mean and variance of pair wise mismatches.

Reduction in effective population size (chapter 3) does not immediately have a large effect on gene diversity but the number of alleles is strongly affected because the rare alleles are lost. The test of Luikart and Cornuet (1998) compares gene diversity and allele number and it can be used to detect recent reductions in effective population size. Such an analysis may be clouded by migration and admixtures.

2.6. Coalescence process

When a wide study of haplotypes (over adjacent loci) from a set of populations is carried out and the ancestry of the haplotypes is traced back in time, a common origin

Box 4.2. The basic coalescence.

The coalescence is a powerful modelling approach in analysing population genetics data. The thinking is reversed. Instead of going forward in time as in the genealogical approach, we go back in time and trace alleles from offspring to parents, further to grand-parents and so on, until a single most recent common ancestor is found. In the figure below we see that the three sampled alleles are descending from another three alleles in generation 9 and these three from two alleles in generation 8. There is coalescence because two alleles in generation 9 are descending from the same ancestral allele in generation 8. If we continue up in the figure there is a point in time (generation 1) until only one ancestral allele remains.

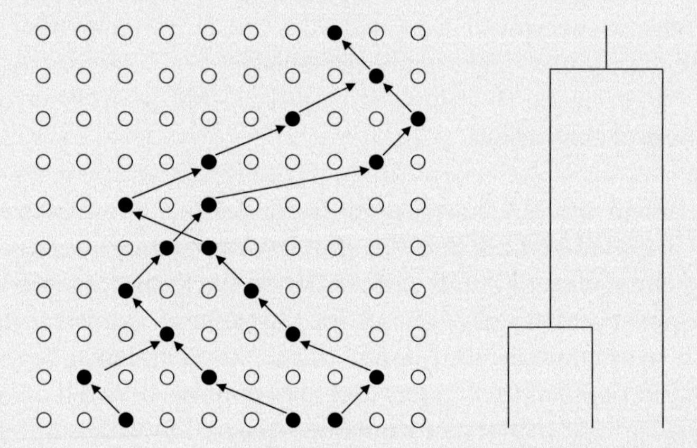

The important point is that coalescence allows to extract genealogical information from DNA data in a computationally efficient way, because it consists of simulating only the sampled genealogies, not the entire population. It can be used in modelling and performing statistical test to infer population demographic history (growing or declining populations) and selection.

will be found. When a common origin is encountered for a pair of haplotypes, the number of ancestral lineages is decreased by one and again the common haplotype is paired with a third one to track an even older lineage and so on, finally ending in one common ancestor of the haplotypes. This genealogical process is called coalescence analysis (reviewed by Rosenberg and Nordborg, 2002). The coalescence process can be used in detecting selection where the depth of the genealogy is indicating the type of selection. Positive selection sweeps an adaptive mutation to fixation leaving behind a shallow star-like genealogy and an excess of low-frequency haplotypes (coupled with low π giving a negative value for Tajima's test) connected to a common ancestor with similar short branches. By contrast, balancing selection results in deep genealogies in which haplotype variants are found at intermediate frequencies (with hitch-hiking variation at linked loci, and consequently a positive value for Tajima's test statistic).

2.7. Establishing the hidden structure of a metapopulation by clustering methods

Phylogenetic techniques based on genetic distances have been the method of choice to assess the genetic diversity of livestock breeds (chapter 5). The approach relies on the *a priori* definition of populations and presents several problems. First, genetic variation within populations is completely ignored. Second, construction of trees using admixed populations, as often happens in livestock, contradicts with the principles of phylogeny reconstruction. And third, it fails to take into account the fact that genetic distances vary greatly according to the marker used and the recent demographic history of the breed (e.g. whether it has passed through a population bottleneck) (Toro and Caballero, 2005).

Recent methods have been developed as a more flexible alternative to genetic distances. The new methods try to divide the total sample of genotypes of a population into an *unknown* number of subpopulations (clusters). This allows the population structure or subdivision to be more flexibly inferred from the data. The clustering methods will separate a set of individuals into several populations when their genetic origin is unknown or to study the correspondence between inferred genetic clusters and known pre-defined population categorisations (like breeds) (Pritchard *et al.*, 2000). The individuals are assigned probabilistically to clusters or jointly to two or more clusters if their genotypes indicate that they are admixed. The methods also estimate, for each individual, the fraction of its genome that belongs to each cluster without any prior information on the structure of the population. Thus, these methods allow to cluster data (genetic mixture analysis) either at group level or at individual level, and also to perform admixture analysis, in which the genome of an individual represents a mixture of alleles of different ancestries.

The algorithms are based on multi-locus genotypes and solved adopting a Bayesian approach computed using Monte Carlo Markov Chain methods and assuming multi-locus genotypes in Hardy-Weinberg and linkage equilibrium within each randomly mating subpopulation. The procedure consists of simultaneously fitting the allele frequencies and assigning individuals to the populations (where some individuals may descent from more than one population). This complex calculation is carried out numerically using a Monte Carlo Markov Chain (MCMC) approach. Three programs are available until now: STRUCTURE (Pritchard *et al.*, 2000), PARTITION (Dawson and Belkhir, 2001) and BAPS (Corander *et al.*, 2003) and their differences are summarised by Pearse and Crandall (2004).

This approach was first applied by Rosenberg *et al.* (2001) in 20 chicken breeds and later it has been applied to pigs, cattle, sheep, goat, dogs and horses. Rosenberg *et al.* (2001) argued that genetically distinctive populations can be identified on the basis of the lack of difficulties encountered in separating them from others. When some populations can easily be separated with only a small number of markers, this could indicate the presence of distinctive multi-locus genetic combinations in those populations. Therefore they suggest that the relative number of loci required for the 'correct' clustering of several populations can be used as a way of identifying those that are genetically distinctive with respect to a collection. In an example in Box 4.3., only six independent loci are enough to separate the two more distinct strains.

3. Utilisation of breed differences

This section is devoted to implementation of the knowledge on breed differences in genetic improvement schemes. The emphasis is on the utilisation of genomic tools for the prediction of heterosis in crosses, for detection of interesting genes in different situations, for introgressing useful genes from one breed to another and for assigning individuals and products to breeds.

3.1. Prediction of heterosis using genetic distances

It is well know in quantitative genetics that crosses between genetically different breeds present hybrid vigour or heterosis especially for traits related to fitness. Heterosis requires directional dominance (for the majority of loci the recessive allele has an unfavourable effect) and differences in frequencies between the lines used in crossing. However, crosses do not always enhance fitness. Crosses between very distant populations may fail to show heterosis and may suffer a reduction in fitness in the F2 generation (*recombination loss*) usually attributed to a break-down of a co-adapted gene complex of favourable epistatic interactions (Dickerson, 1969).

Box 4.3. An example on the use of clustering methods.

The analysis comprises five strains of the Iberian breed and the Duroc breed (Fabuel *et al.*, 2004). The strains are classified in two clusters by the STRUCTURE algorithm of Pritchard *et al.* (2000), with all the Iberian strains falling into one cluster, and the Duroc breed constituting the other one. Torbiscal and Guadyerbas strains are the populations whose genomes are differentiated most unambiguously from the Duroc.

Table 4.2 presents also the results when the cluster analysis is carried out only in the Iberian pig population and assuming the same number (five) of clusters (as the number of predefined strains is usually considered). On average, 98.5% of the Torbiscal genomes and 99.5% of the Guadyerbas genomes are classified as two separate clusters. However, the results are less clear for the other populations, whose genomes are attributed to diverse clusters. The last analysis also emphasises that the first two strains constitute more defined populations than the others.

Table 4.2. Proportion of membership of each predefined population after assuming either two or five possible clusters for the chosen set of breeds.

	two clusters assumed		five clusters assumed				
population	1	2	1	2	3	4	5
Torbiscal	0.001	0.999	0.004	0.003	0.002	0.985	0.006
Guadyerbas	0.001	0.999	0.001	0.001	0.001	0.002	0.995
Retinto	0.011	0.989	0.449	0.451	0.009	0.084	0.007
Entrepelado	0.050	0.950	0.527	0.419	0.008	0.030	0.016
Lampiño	0.010	0.990	0.321	0.223	0.351	0.024	0.081
Duroc	0.997	0.003					

Because heterosis is proportional to the differences in gene frequencies in the parental lines, it has been suggested that it would be possible to make marker-based predictions of hybrid performance based on genetic distances, despite having only indirect estimates of allele frequencies for the interesting traits *via* the anonymous markers spread through the genome. This prediction is efficient in some crop plants when the lines included in the cross are related by pedigree or can be traced back to common ancestral populations. But it is not efficient if the lines are unrelated or originating from different populations because the associations between marker and trait loci are not the same in different populations. In domestic animals the exercise has been done in chickens (Gavora *et al.*, 1996), who found a high significant correlation (0.68-0.87) between band sharing for DNA fingerprints and egg production traits. On the other hand, if genes important for a trait are known together with the dominance and epistatic interactions, it will be possible to make appropriate crosses such that the most desirable genotype is produced.

3.2. QTL detection

QTL (Quantitative Trait Locus) is a genetic region that affects phenotypic variation. The abundance of molecular information allows us to identify the specific regions of the genome that affect phenotypes of interest. According to a QTL data base for domestic animals, in September 2006 the number of QTL's detected was 1287, 630 and 657 in pigs, cattle and chicken, respectively. It is important to recognise that a substantial number of them have been located by profiting from the existence of divergent breeds which have been be used in crossbreeding designs.

Genomic research facilitates localising and characterising genes and their function. Breeds with extreme phenotypes have proven to be very fruitful, especially for traits where a major part of the variation can be linked to one or two genes. There are several such examples and the list is growing: Chinese and European pig breeds that differ in prolificacy, wild boar and commercial pig breeds differing in growth and fatness, standard and double muscling cattle breeds and so on. It is probable that, in many cases, the greatest benefits from QTL mapping and localising the actual gene are in understanding the nature of genetic variation and the way how genes will function and interplay with each other, rather than in enhancing conventional breeding programmes with marker or gene assisted selection.

3.2.1. Candidate genes

There are two main strategies to locate genes that affect traits of interest. The first is the candidate gene approach that focuses on a few known genes whose physiological function suggests that different alleles could be responsible for differences in the phenotypic values of the trait. Obviously, to identify polymorphism within the candidate gene, advantage can be taken by concentrating on breeds that are widely diverged for the trait of interest. For example, in a pioneering study by Rothschild *et al.* (1996) in pigs, the focus was on the oestrogen receptor gene (ESR) because of its positive impact on reproduction. They identified polymorphism in the Large White breed and in Chinese Meishan pigs, an extremely prolific breed from China, and in their composite crossbreds. There were three genotypes AA, AB and BB and the association studies showed that the difference between AA and BB genotypes was 2.3 pigs per litter in the Meishan composite population and 0.9 in the Large White. Unfortunately, more recent studies have questioned these results indicating that the polymorphism is merely a marker and not the causative mutation (Alfonso, 2005).

The most thrilling examples so far are on the genes related to muscle growth and meat production. E.g. the gene product *myostatin* acts as a negative regulator of skeletal muscle growth and double-muscled cattle are homozygous for loss-of-function mutations. Grobet *et al.* (1997) reported five different mutation alleles present if different breeds of beef cattle. The *callipyge* gene in sheep causes muscular hypertrophy. The mutation is very recent and it has been mapped on sheep chromosome 18. The trait has been shown to have peculiar inheritance – called polar overdominance – where the phenotype is expressed only in heterozygotes and only when the mutant allele has been inherited from the sire (Cockett *et al.*, 1996).

An even more exciting finding has been made in the meaty Texel sheep (Clop *et al.*, 2006). First, a QTL analysis was carried out in Romanov x Texel F2 generation. The QTL for muscularity was mapped on chromosome 2 and was further confirmed to be the *myostatin* gene, a good candidate for the trait. Surprisingly this turned out not to be in the coding region of the gene, as was the case with double muscling in cattle, as it was discovered that the mutation effect is mediated by microRNAs (miRNAs) that regulate the expression of other genes. Animal miRNAs usually have complementary sites in the 3'UTR (untranslated region) of mRNAs. The annealing of the miRNAs to the mRNA affects the protein translation. A mutation in *myostatin* creates an illegitimate target site for at least two miRNA-mediated traslational downregulations and reductions in myostatin concentrations contributing to muscular hypertrophy. The same group is showing how mammal genomes have large numbers of mutations creating or destroying miRNA candidate sites (Clop *et al.*, 2006). These may serve as important factors in regulating variation in quantitative traits.

In conclusion, the gene findings can be used in enlightening us on the complexities of gene expression and control. This information can at its best used in formulating feeding or adjusting the production environment to match with the genotypes. A good example on this would be fat metabolism (e.g. Lock and Bauman, 2004). We also learn all the time more about the difficulties and prospects for modifying the genes themselves (gene transfer or knockout) or their expression (knockdown).

3.2.2. Genome scans of crossbred populations

The second strategy for detecting QTL's is called the genome scan. The idea is to screen the entire genome for regions that are associated with variation in the traits of interest, whether or not those regions are known to contain potential candidate genes. About 100-150 of evenly spaced polymorphic markers are chosen to cover the whole genome. Markers have no economic value themselves but the objective in these genome screens is to find markers associated with QTL's. This will require polymorphism in both the markers and the QTL. The markers should be close to QTL's and finally no independent segregation should occur between markers and the QTL (but instead co-segregation or linkage disequilibrium).

The most powerful and easiest design to detect QTL's by a genome scan is a backcross or an F2 between two divergent strains. In this case, substantial polymorphism is expected and a crossed population harbours high linkage disequilibrium. Furthermore, the disequilibrium between a marker and the QTL is proportional to the genetic distance in such a way that finding a marker with effect on the trait implies that the marker is located in a region close to the QTL affecting the trait.

The first major genome scan utilising breed differences was reported in 1994 in pigs from a cross between Large White and its wild ancestor, the wild boar (Andersson *et al.*, 1994). Large QTL effects were found influencing both growth and fatness; the OTL's were located on chromosome 4. After that study, several genome scans have been developed involving the typical commercial breeds and more special breeds such as Meishan, Iberian, Berkshire, Magalitza or wild boar.

As another example, in a more recent study (Carlborg *et al.*, 2003) a large intercross comprising 851 F2 individuals between the domestic White Leghorn chicken and the wild ancestor, the Red Jungle Fowl, was used in a QTL study for growth and egg production. The QTL analysis of growth traits revealed 13 loci that showed genome-wide significance and the four major growth QTL's explained 50 and 80% of the difference in adult body weight between the founder populations for females and males, respectively.

The cross between divergent lines is very powerful to detect QTL's because: (1) the environment in a small experiment can be controlled, (2) traits that are not routinely scored can be studied in a detailed production trial (hormonal levels, immunological parameters, quality traits) and, (3) the statistical power is high. For example a QTL responsible causing a difference of one phenotypic standard deviation between the parental lines will be detected with a probability greater that 90% in an F2 design scoring only 200-300 individual.

However, the F2 design has some disadvantages because the bracket for placing the QTL may be very long, about 20 cM. With such a resolution, it is difficult to pursue the study to find the causal mutation, even if the linkage equilibrium in the parental lines is favourable. For more recombinants and shorter confidence intervals, intercrosses F3, F4, F5, etc should be produced, which delays finding the actual gene. An alternative is

Box 4.4. Detecting QTL's in a F2 design.

Let us imagine that we have genotyped animals from two different breeds. Morphologically one breed has a large body size and a white colour and the other a small body size and a black colour. Let us consider a marker with a different allele being fixed in the lines. In the F2 animals all possible combinations of morphological and marker alleles will appear. We can classify the animals to the genotypes MM, Mm and mm. The MM animals have a large body size, whilst the mm animals are smaller with the Mm being intermediate. This would suggest that the M/m marker is close to a QTL related to growth. This is not the case for the colour trait because now all the colour phenotypes appear with the same frequencies in the different weight classes. Obviously, sophisticated statistical techniques (regression, maximum likelihood and Bayesian methods) will be implemented in practice in estimating both the position and the effect of the putative QTL

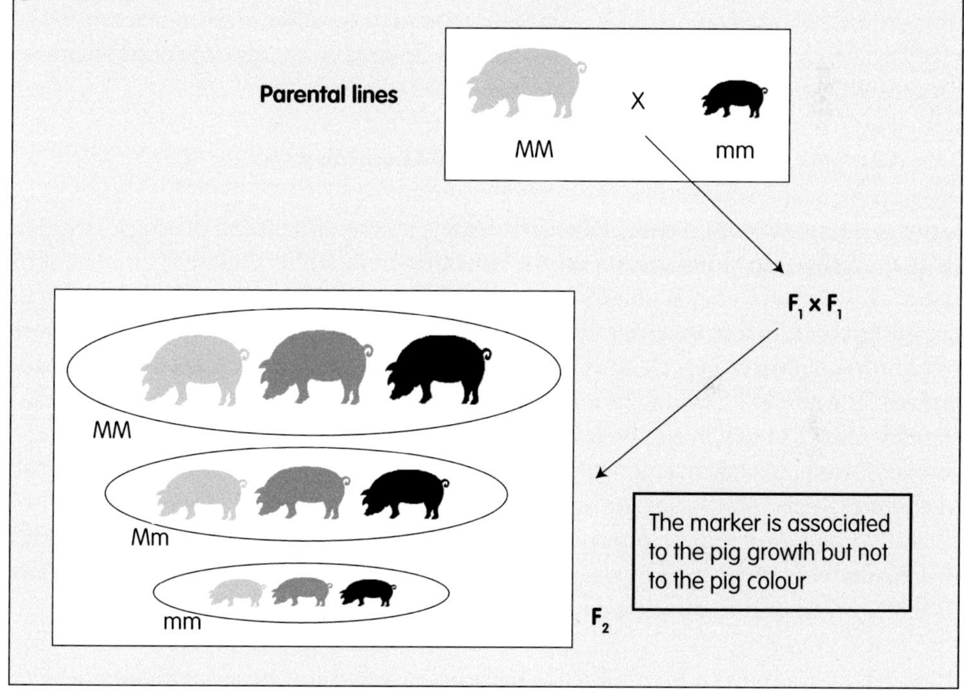

to use a synthetic breed from the cross of two divergent breeds and randomly interbred for several generations (cf. admixture mapping discussed later).

3.2.3. Selection mapping of QTL

In analysing large comparative genomic data sets, e.g. large SNP data sets it is possible to determine how and where both positive and negative selection has affected variation in farm animal populations. When a candidate gene for a trait, included in the breeding programme, is associated with a selective sweep in the genome, this could be used as evidence for its influence on the genetic variation. Myostatin was described before as the mutations causing extreme muscle mass (double muscling) of Belgian Blue (Charlier *et al.*, 1995). Double muscling has been actively selected in several cattle breeds - despite its adverse effects on calving. When three double-muscled and six non-double-muscled cattle breeds were analysed for the gene and 18 markers on the same chromosome, there was a correlation between heterozygosity and the distance from the gene and linkage disequilibrium was greater in double-muscled breeds (Wiener *et al.*, 2003). The results were not as consistent as would be expected and the authors considered the age and history of selection to be distorting the patterns. Analogously, before genetic diversity patterns are implemented in locating QTL mediating variation in production traits, it will be important to have information on such possible deviating factors in the breeds. Selection mapping has been successfully applied in comparing rat populations for resistance to anticoagulants (Kohn *et al.*, 2000).

3.2.4. QTL mapping in bottlenecked and admixture populations

Linkage disequilibrium is crucial for associating markers with loci mediating a variation in quantitative trait. Information on the distribution of LD is therefore important in assessing the results from genome wide QTL studies. Such information is also critical in designing cost-efficient investigations. Many breeds are stemming from a small number of founders and have experienced bottlenecks alternating with periods of population growth. The number of founders and bottlenecks affect the sampling of haplotypes and thereby cause LD which can be used in localising genes in the QTL regions. Linkage disequilibrium extends over longer distances in young populations and enables the use of fewer markers in an association study, but with the drawback that the position of the causal mutation will be poorly defined. Within modern dairy breeds LD extends over several tens of centi-Morgans (Farnir *et al.*, 2000). The older haplotypes shared by different breeds can then be used for a higher resolution mapping.

There has been varying degrees of gene exchange between farm animal populations since domestication, and where possible, even introgression of genes from wild populations. The formation of a breed has in some cases aimed at the utilisation of more developed breeds in upgrading local landrace populations to reach a generally accepted breed status. A good example of this is the wide use of Shorthorn in the development of several European cattle breeds.

Modern genetic improvement programmes in pigs and chicken are based on lines specialised on a small number of traits, whilst the production animals are their hybrids.

The lines – usually termed pure lines - are originating from planned crosses of few conventional breeds with desirable traits. Today the formation of a new population by interbreeding genetically divergent parental populations is called admixture. The gene flow that takes place during admixture results in a temporary generation of long haplotype blocks. This will facilitate efficient genome wide admixture mapping (Flint *et al.*, 2005) of QTL's over few generations before the linkage disequilibrium is broken by recombination. It has a moderate mapping resolution and serves as a cost-efficient approach for the initial genome scan. Admixture mapping is obviously efficient in traits for which the parental populations have a large phenotypic difference.

3.3. Introgression

Sometimes a breed (donor), otherwise unproductive, could have a target gene that will be of interest to introduce into an economically important breed (recipient). This is usually done by an introgression programme. It consists of forming an initial cross between the breeds followed by repeated backcrosses to the recipient line to recover the economically important genome. The target gene is maintained in the backcross generation through selection of donor gene carriers. After some generation of backcrossing the programme will finish by a generation of intercrossing to make the population homozygous for the desired allele

Genetic markers could be useful in introgression programmes in two ways (Dekkers and Hospital, 2002). First, markers can be used to select individuals at each backcross generation which are heterozygous for the desired allele or homozygous in the last generation of intercross (foreground selection). Secondly, markers can be used to enhance the recovery of the recipient genome (background selection). When the donor and recipient breeds are initially crossed, new variation is also introduced into the recipient population and the effective population size is increased. In recovering the original recipient genome, much attention should be paid to avoiding losses of variation and to utilising the new variation. This would require a large population and data collection on relevant traits.

The introgression strategy is the main genetic improvement method in plants but there are still very few applications in domestic animals. An example is the naked neck gene in chicken which reduces plumage in chicken and makes animals more tolerant to heat. It was introgressed from low body weight landrace chickens into a commercial meat-type Cornish chicken (Yancovich *et al.*,1996). Other candidates would be the Booroola gene in sheep (Montgomery *et al.*, 1992; Campbell *et al.*, 2003), the estrogen receptor in pigs (Rotschild *et al.*, 1996) and the polled allele in cattle (Drogemuller *et al.*, 2005).

Introgression of several genes is more complicated, because the selection of animals carrying all the desired alleles in the backcross generation would require a very large population size. The alternative strategy is to use a pyramid design. The genes are introgressed one by one in different lines and the lines are crossed afterwards. There is an important project at the International Livestock Research Institute (ILRI) in Kenya, aimed to introduce three QTL's that confer resistance to trypanosomiasis (equivalent to human sleeping sickness, transmitted by a tsetse fly), from N'Dama resistant cattle into highly productive susceptible breeds. The most cost-efficient strategy seems to be to introgress two QTL's in one line and two QTL's in the other (in such a way that there will be one QTL in common) and to make the appropriate final crosses. Such strategy has shown to be successful in an experimental introgression of trypanotolerance QTL's in mice (Koudandé *et al.*, 2005).

3.4. Assignment of animals and products to breeds

Because breeds are different at the genetic level, genetic markers can provide a tool to identify the breed to which an animal belongs to, when genealogical information is absent. Breed conservation usually emphasises the maintenance of a pure breed and in this context molecular markers can detect if introgression of crossbreeding has occurred. The assignment tests are also relevant to animal traceability from birth to market that is increasingly requested as an element of a food safety assurance system. Both producers and consumers are interested in the development of reliable methods of traceability for animals and their products. Furthermore, the marketing link between product and breed can improve the profitability of local breeds (chapter 2).

There are two ways to tackle the assignment problems with molecular genetics. In the more classical approach (often called supervised method) the task is to assign an anonymous sample to one of several reference populations. The populations are given and the marker allele frequencies are supposed to be known from some previous studies. The methods are simple and the computations are rapid; the only drawback is that the reference populations must be carefully identified and characterised with an adequate number of individuals.

The simplest approach to the supervised methods has been developed by Blott *et al.* (1999). The individual x is assigned to breed k rather than to breed j if

$$q_j f_j(x) \leq q_k f_k(x)$$

where q_j and q_k are the prior probabilities of drawing an individual from breed j or breed k, (some breeds are more numerous than others), and $f_j(x)$ and $f_k(x)$ are the probabilities that the genotype x occurs in the breeds j and k, respectively. Under Hardy-Weinberg equilibrium at the locus these probabilities are simply p^2_{jx} for the homozygotes and $2p_{jx}p_{jy}$ for the heterozygotes. The genotype probabilities in the crossbred animal are $p_{jx}p_{jy}$ for the homozygotes and $p_{jx}p_{ky}+p_{jy}p_{kx}$ for the heterozygotes where p_{jx} and p_{kx} are the frequencies of allele x in breeds j and k and p_{jy} and p_{ky} being the same for allele y.

There are two types of errors. The type I error, or the proportion of individuals of one breed that are allocated to another breed, and the type II error, or the proportion of individuals that are allocated to one breed but are really of another breed. It is found that an error rate < 5% requires 11-18 microsatellites (or about 65-100 SNP's). The most discriminatory markers are those with the highest expected heterozygosity and number of alleles and obviously the most powerful will be those that are fixed for different alleles in the different breeds (private alleles, chapter 3). The above approach has been applied in several settings: cattle breeds (Cañón *et al.*, 2000) or horses (Bjørnstad and Røed, 2001). The most widely used supervised method is the one implemented in GeneClass2 available at http://www.ensam.inra.fr/URLB.

The unsupervised methods correspond to the clustering methods (cf. STRUCTURE, see paragraph 2.7 and Box 4.3) that allow also to assign individuals to populations or mixtures. The advantage of clustering is that it makes possible to take into account complex genetic situations, such as admixtures. However, besides of being computationally demanding, when the number of populations is left as a parameter to be estimated, there are situations that the produced clusters do not represent real populations in the field.

4. Conclusions

The genomic research is a powerful approach to analyse the history of animal domestication. It is clear that mtDNA and autosomal markers (microsatellites and SNP's) are very useful tools in such analyses. Often a joint use of different types of markers is needed to integrate different studies and to complete the historical picture. The important male-mediated gene flow can be included in the studies using variable Y chromosome markers. Hence we are able to simultaneously detect the influence of male and female lineages and selection.

Until recently the population genetic analyses have been carried out using neutral markers. The information on loci affecting the variation in economically important traits or in fitness is accruing. The use of such loci in diversity research can illuminate the spread of favoured alleles and the modification of genomes along the history of populations. Another recent important feature in genomics, is the availability of high

Box 4.5. Traceability in Iberian pigs.

The production of Iberian pigs represents only 5% of the whole Spanish pork production but it has a great significance. It feeds a processing industry that is the source of the most prestigious dry-cured products. The optimum quality is associated with purebred Iberian genotypes which are commonly crossbred with other breeds, mainly Duroc, to improve the lean content of the carcasses. An essential part of this industry is interested in building a link between products and purebred Iberian animals to further distinguish their products within a heterogeneous market. To avoid possible fakes in the labelling of meat and cured products, identifying purebred and crossbred genotypes decreases the risk of introgression of foreign genes into the Iberian genetic pool.

Several techniques to discriminate the genotypes of purebred and crossbred Iberian pigs have been carried out. A panel consisting of the nine most frequent markers allows the discrimination between purebred and crossbred animals (Alves *et al.*, 2002). The probabilities of exclusion of the pure Iberian origin were 0.97 and 0.78, for crossbred individuals with 50 or 25% of Duroc genes, respectively. García *et al.* (2006) used both supervised and unsupervised procedures and found that the proposed procedure detected up to 20% of commercial ham samples with a genetic composition incompatible with present legislation – either because the Duroc genome was present in a percentage greater than that permitted, or because of the significant presence (>25%) of white coat pig genomes. They also showed that the probability of finding an illegal cured ham was greater in restaurants than in retail grocery stores, and in medium-low category restaurants or stores than in higher category establishments.

through-put systems which would open up possibilities for high resolution analyses of the genome diversity. Nevertheless the choice of markers remains important and needs careful planning in terms of power and costs.

The analyses have given support for a surprisingly high number of domestication events. In the history of farm animal populations, we can detect the traces of genetic drift, selection, introgression, admixture and pinpoint roughly the occurrence and time points of population expansions and contractions.

The understanding of breed differentiation is offering new material and tools that can immediately be used in genetic improvement programmes. The markers which are best in quantifying the variation between breeds can be used in assigning individuals and products to breeds. The loci which are causing the most substantial deviations between breeds can be easily found and may be useful in upgrading the low performing breeds, and again genomics can be used in accelerating and sharpening the introgression of useful alleles genes from one population to another.

References

Adalsteinsson, S. 1981. Origin and conservation of farm animal populations in Iceland. Zeitschrift für Tierzüchtung und Züchtungsbiologie 98: 258-264.

Akey, J.M., G. Zhang, K. Zhang, L. Jin and M.D. Shriver., 2002 Interrogating a high-density SNP map for signatures of natural selection. Genome Research 12: 1805–1814.

Alfonso, L. 2005. Use of meta-analysis to combine candidate gene association studies: application to study the relationship between the ESR PvuII polymorphism and sow litter size. Genetics Selection Evolution 37: 417-435

Alves, E., C. Castellanos, C. Ovilo, L. Silió and C. Rodríguez, 2002. Differentiation of the raw material of the Iberian pig meat industry based on the use of amplified fragment length polymorphism. Meat Science 61:157-162.

Andersson, L. and M. Georges., 2004. Domestic animal genomics: deciphering the genetics of complex traits. Nature Reviews Genetics 5: 202-212.

Andersson, L., C.S. Haley, H. Ellegren, S.A. Knott, M. Johansson, K. Andersson, L. Andersson-Eklund, I. Edfors-Lilja, M. Fredholm, I. Hansson, J. Hakansson and K. Lundstrom., 1994. Genetic mapping of quantitative traits loci for growth and fatness in pigs. Science 263: 1771-1774.

Beaumont, M. A., W. Zhang and D.J. Balding., 2002 Approximate Bayesian computation in population genetics. Genetics 162: 2025–2035.

Beja-Pereira A., G. Luikart, P.R. England, D.G. Bradley, O.C. Jann, A.T. Chamberlain, T.P. Nunes, G. Bertorelli, S. Metodiev, N. Ferrand and G. Erhardt., 2003 Milk Drinkers: Gene-culture coevolution between cattle and humans. Nature Genetics 35: 311-313.

Beja-Pereira, A., P.R. England, N. Ferrand, S. Jordan, A.O. Backhiet, M.A. Abdalla, M. Mashkour, J. Jordana, P. Taberlet and G. Luikart, G. 2004. African origins of the domestic donkey. Science 304: 1781.

Bjørnstad, G. and K.H. Røed., 2001. Breed demarcation and potental for breed allocation of horses assessed by microsatellite markers. Animal Genetics 32: 59–65.

Blott, S.C., J.L. Williams and C.S. Haley., 1999. Discrimination among cattle breeds using genetic markers. Heredity 82: 613-619.

Bruford, M.W., D.G. Bradley and G. Luikart., 2003 *DNA* markers reveal the complexity of livestock domestication. Nature Reviews Genetics 4: 901-910.

Campbell B.K., D.T. Baird, C.J. Souza and R. Webb, 2003. The FecB (Booroola) gene acts at the ovary: in vivo evidence. Reproduction 126: 101-111.

Cañón, J., P. Alexandrino, A. Beja-Pereira, I. Bessa, C. Carleos, Y. Carretero, S. Dunner, N. Ferrand, D. García, J. Jordana, D. Laloë, A. Sánchez and K. Moazami-Goudarzi., 2000. Genetic diversity of European local beef cattle breeds for conservation purposes. Genetics Selection Evolution 33: 311-332.

Carlborg, Ö., S. Kerje, K. Schutz, L. Jacobsson, P. Jensen and L.A. Andersson., 2003 A Global Search Reveals Epistatic Interaction between QTL's for Early Growth in the Chicken. Genome Research 13: 413-421.

Charlesworth, B., M.T. Morgan and D. Charlesworth, D. 1993 The effect of deleterious mutations on neutral molecular variation. Genetics 134: 1289–1303.

Charlier, C., W. Coppieters, F. Farnir, L. Grobet, P.L. Leroy, C. Michaux, M. Mni, A. Schwers, P. Vanmanshoven, R. Hanset and M. Georges, 1995. The *mh* gene causing double-muscling in cattle maps to bovine chromosome 2. Mammalian Genome 6: 788–792.

Clop, A., F. Marcq, H. Takeda, D. Pirottin, X. Tordoir, B. Bibe, J. Bouix, F. Caiment, J.M. Elsen, F. Eychenne, C. Larzul, E. Laville, F. Meish, D. Milenkovic, J. Tobin, C. Charlier and M. Georges, 2006. A mutation creating a potential illegitimate microRNA target site in the myostatin gene affects muscularity in sheep. Nature Genetics. 38: 813-818.

Cockett, N.E., S.P. Jackson, T.L. Shay, F. Farnir, S. Berghmans, G.D. Snowder, D.M. Nielsen, and M. Georges, 1996. Polar overdominance at the ovine callipyge locus. Science 273: 236–238.

Corander, J., P. Waldmann and M.J. Sillanpaa., 2003 Bayesian analysis of genetic differentiation between populations. Genetics 163: 367-374.

Daly, M. J., J.D. Rioux, S.E. Schaffner, T.J. Hudson and E.S. Lander., 2001. High-resolution haplotype structure in the human genome. Nature Genetics 29, 229–232.

Darwin, C., 1868. The variation of animals and plants under domestication. John Murray. London.

Dawson, K.J. and K. Belkhir, 2001. A bayesian approach to the identification of panmictic populations and the assignment of individuals. Genetical Research 78: 59-73.

Diamond, J., 2002. Evolution, consequences and future of plant and animal domestication. Nature 418: 700-707.

Dickerson, G.E., 1969 Experimental approaches in utilizing breed resources. Animal Breeding Abstracts 37: 191-202.

Dekkers, J.C.M. and F. Hospital., 2002. The use of molecular genetics in the improvement of agricultural populations. Nature Reviews Genetics 3: 22-32.

Dobney, K. and G. Larson, 2006. Genetics and animal domestication: new windows on an elusive process. Journal of Zoology 269: 261-271.

Drogemuller, C., A. Wohlke, S. Momke and O. Distl, 2005. Fine mapping of the polled locus to a 1-Mb region on bovine chromosome 1q12. Mammalian Genome 16: 613-620.

Fabuel, E., C. Barragán, C-L., Silio´, M.C. Rodriguez and M.A. Toro, 2004. Analysis of genetic diversity and conservation priorities in Iberian pigs based on microsatellite markers. Heredity 93: 104-113.

Farnir, F., W. Coppieters, J.J. Arranz, P. Berzi, N. Cambisano, B. Grisart, L. Karim, F. Marcq, L. Moreau, M. Mni, C. Nezer, P. Simon, P. Vanmanshoven, D. Wagenaar and M. Georges, 2000. Extensive Genome-wide Linkage Disequilibrium in Cattle. Genome Research 10: 220-227.

Flint, J., W. Valdar, S. Shifman and R. Mott., 2005. Strategies for mapping and cloning quantitative trait genes in rodents. Nature Reviews Genetics 6: 271-286.

Galton, F., 1883. Inquiries into Human Faculty and its Development. Macmillan, New York.

García, D., A. Martínez, S. Dunner, J.L. Vega-Pla, C. Fernández, J.V. Delgado and J. Cañon, 2006. Estimation of the genetic admixture composition of Iberian dry-cured ham samples using DNA multilocus genotypes. Meat Science 76: 560-566.

Gavora, J.S., R.W. Fairfull, B.F. Benkel, W.J. Cantwell and J.R. Chambers., 1996. Prediction of heterosis from DNA fingerprints in chicken. Genetics 144: 777-784.

Grobet L., L.J.R. Martin, D. Poncelet, D. Pirottin, B. Brouwers, J. Riquet, A. Schoeberlein, S. Dunner, F. Menissier, J. Massabanda, R. Fries, R. Hanset and M. Georges, 1997. A deletion in the bovine myostatin gene causes the double-muscled phenotype in cattle. Nature Genetics. 17: 71-74.

Hillier, L.W., W. Miller, E. Birney, *et al.* (International Chicken Genome Sequencing Consortium), 2004. Sequence and comparative analysis of the chicken genome provide unique perspectives on vertebrate evolution. Nature 432: 695-716.

Johnson, G.C.L., L. Esposito, B.J. Barratt, A.N. Smith, J. Heward, G. Genova, H. Ueda, H.J. Cordell, I. Eaves, F. Dudbridge, R.C.J. Twells, F. Payne, W. Hughes, S. Nutland, H. Stevens, P. Carr, E. Tuomilehto-Wolf, J. Tuomilehto, S.C.L. Gough, D.G. Clayton and J.A. Todd. 2001. Haplotype tagging for the identification of common disease genes. Nature Genetics 29: 233–237.

Kohn, M.H., H.J. Pelz and R.K. Wayne., 2000. Natural selection mapping of the warfarin-resistance gene. Proceedings of the National Academy of Sciences of the USA 97: 7911-7915.

Koudandé, O.D., J.A.M. van Arendonk and F. Iraqi, 2005. Marker-assisted introgression of trypanotolerance QTL in mice. Mammalian Genome 16: 112-119.

Larson, G., K. Dobney, U. Albarella, M. Fang, Matisoo-Smith, J. Robins, S. Lowden, H. Finlayson, T. Brand, E. Willerslev, P. Rowley-Conwy, L. Andersson and A. Cooper, 2005. Worldwide phylogeography of wild boar reveals multiple centers of pig domestication. Science 11: 1618-1621.

Lewontin, R.C. and J.K. Krakauer, 1973. Distribution of gene frequency as a test of the theory of the selective neutrality of polymorphisms. Genetics 74: 175–195.

Lindgren, G., N. Backström, J. Swinburne, L. Hellborg, A. Einarsson, K. Sandberg, G. Cothran, C. Vilà, M. Binns and H. Ellegren, 2004. Limited number of patrilines in horse domestication. Nature Genetics 36: 335-336.

Lock, A.L. and D.E. Bauman., 2004 Modifying milk fat composition of dairy cows to enhance fatty acids beneficial to human heath. Lipids 39: 1197-1206.

Loftus, R.T., D.E. MacHugh, D.G. Bradley, P.M. Sharp and E.P. Cunningham., 1994 Evidence for two independent domestications of cattle. Proceedings of the National Academy of Sciences 91: 2757-2761.

Luikart, G. and J.M. Cornuet, 1998 Empirical evaluation of a test for detecting recent historical population bottlenecks. Conservation Biology 12: 228–237.

Maynard-Smith, J. and J. Haigh, 1974. The hitch-hiking effect of a favourable gene. Genetical Research 23: 23–35.

Meadows, J. R. S., R.J. Hawken and J.W. Kijas, 2004. Nucleotide diversity on the ovine Y chromosome. Animal Genetics 35: 379-385.

Mignon-Grasteau, S., A. Boissy, J. Bouix, J.M. Faure, A.D. Fisher, G.N. Hinch, P. Jensen, P. Le Neindre, P. Mormede, P. Prunet, M. Vandeputte and C. Beaumont 2005. Genetics of adaptation and domestication in livestock. Livestock Production Science 9: 3-14.

Montgomery, G.W., K.P McNatty and G.H. Davis., 1992 Physiology and moceluar genetics of mutations that increase ovulation rate in sheep. Endocrine Reviews. 13: 309-328.

Pearse, D.E. and K. Crandall, 2004. Beyond Fst: Analysis of population genetic data for conservation. Conservation Genetics 5: 585-602.

Pritchard, J.K., M. Stephens and P.J Donnelly, 2000. Inference of population structure using multilocus genotype data. Genetics 155: 945-959.

Rogers, A.R. and H. Harpending, 1992. Population growth makes waves in the distribution of pairwise genetic differences. Molecular Biology and Evolution 9: 552–569.

Rosenberg, N.A. and M. Nordborg, 2002. Genealogical trees, coalescent theory and the analysis of genetic polymorphisms. Nature Reviews Genetics. 3: 380-390.

Rosenberg, N.A., T. Burke, K. Elo, M.W. Feldman, P. Friedlin, M.A.M. Groenen, J. Hillel, A. Mäki-Tanila, M. Tixier-Boichard, A. Vignal, K. Wimmers and S. Weigend, 2001. Empirical evaluation of genetic clustering methods using multilocus genotypes from 20 chicken breeds. Genetics 159: 699-713.

Rothschild, M., C. Jacobson, D. Vaske, C. Tuggle, L. Wang, T. Short, G. Eckardt, S. Sasaki, A. Vincent, D. McLaren, O. Southwood, H. van der Steen, A. Mileham and G. Plastow, 1996. The estrogen receptor locus is associated with a major gene influencing litter size in pigs. Genetics 93: 201-205.

Tajima, F. 1989. The effect of change in population size on DNA polymorphism. Genetics 123: 597-601.

Tapio, M., N. Marzanov, M. Ozerov, M. Ćinkulov, G. Gonzarenko, T. Kiselyova, M. Murawski, H. Viinalass and J. Kantanen., 2006. Sheep mitochondrial DNA variation in European, Caucasian and Central Asian areas. Molecular Biology and Evolution 23: 1776–1783.

Toro, M.A. and A. Caballero, 2005. Characterisation and conservation of genetic diversity in subdivided populations. Philosophical Transactions of the Royal Society of London B 360: 1367-1378.

Trut, L.M., 1999. Early canid domestication: the farm-fox experiment. American Scientist 87: 160-168.

Wall, J. D. and J.K. Pritchard., 2003 Haplotype blocks and linkage disequilibrium in the human genome. Nature Reviews Genetics. 4: 587-597.

Wiener, P., D. Burton, P. Ajmone-Marsan, S. Dunner, G. Mommens, I.J. Nijman, C. Rodellar, A. Valentini and J.L.Williams, 2003. Signatures of selection? Patterns of microsatellite diversity on a chromosome containing a selected locus. Heredity 90: 350-358.

Yancovich, A., I. Levin, A. Cahaner and J. Hillel, 1996. Introgression of the avian naked neck gene assisted by DNA fingerprint. Animal Genetics 27: 149-155.

Zeder, M.A., D.G. Bradley, E. Emshwiller and B.D. Smith., 2006. Documenting Domestication: New Genetic and Archaeological Paradigms. University of California Press.

Zhang, K., J.M. Akey, N. Wang, M. Xiong, R. Chakraborty and L. Jin., 2003. Randomly distributed crossovers may generate block-like patterns of linkage disequilibrium: an act of genetic drift. Human Genetics 113: 51–59.

Chapter 5. Measuring genetic diversity in farm animals

Herwin Eding[1] *and Jörn Bennewitz*[2]
[1]*Federal Agricultural Research Centre, Institute for Animal Breeding, Höltystraße 10, 31535 Neustadt, Germany*
[2]*Institute of Animal Breeding and Husbandry, Christian-Albrechts-University of Kiel, 24098 Kiel, Germany*

Questions that will be answered in this chapter:

- *Why measure genetic diversity?*
- *What measures of genetic diversity are there and when are they appropriate?*
- *How are these measures related and interpreted?*
- *What is Weitzman diversity, what is Core set diversity and which should one use?*

Summary

In chapter 3 the general meaning of genetic diversity was explained, as well as the basic principles underlying the origin and influence of genetic diversity within and between breeds. In this chapter we will present a number of widely used methods to measure genetic diversity in farm animal species. These range from genetic distances and F-statistics to kinships as methods to estimate genetic diversity between and within populations. In the second part we will discuss two frameworks to summarise these diversity measures: the Weitzman and the Core set method.

1. Introduction

In the last decades much research has been focused on the determination of genetic diversity and the uniqueness of breeds in order to decide on genetic conservation priorities. Many of the measures are based on population genetic theory and are using molecular data. Such measures reveal genetic diversity with neutral genetic markers (chapter 3) and support conservation decisions aiming to maintain the genetic flexibility and potential for changes (chapter 2). While genetic diversity can be defined in a number of ways (in terms of conservation of species or farm animal breeds or single alleles) for this chapter the general objective is maintenance of the genetic variance in the species.

1.1. Why measure genetic diversity?

Within livestock species, the genetic diversity is most obvious as the spectrum and number of breeds. Breeds are defined as populations within a species of which the members can be determined by a set of characteristics particular to the breed (FAO, 1998), although there exist many alternative definitions (chapter 3). The FAO definition assumes that in phenotypes of characteristics or traits, there is a clear boundary between populations. This may be clear in Europe where separation of breeds from others was an intentional process accompanied by the establishment of herd books, some 200 years ago (Ruane, 1999). In other regions, e.g. in Africa, such a clear definition of breeds is not always possible, due to widespread crossing between populations. Assigning animals to breeds in these regions is subjective and often questionable (Scherf, 2000).

We want to have a more general picture of the variation and concentrate on variation in traits or genotypes in populations, whether they are breeds or sub-populations within a livestock species. The genetic variance in a trait can be partitioned using the coefficient of kinship in between and within population components (chapter 3). With molecular genetic techniques, such as genotyping with microsatellite markers, genetic diversity between breeds is mostly studied using genetic distances or genetic similarities and derived quantities (molecular coancestries, marker estimated kinships). Genetic distances express the differences between populations either in terms of numbers of mutations or in terms of differences in allele frequencies or genetic drift. Breed formation occurred recently on the evolutionary scale. For this reason, genetic diversity between populations is usually quantified using measures based on genetic drift only and ignoring the effect of mutation. Within a breed the diversity is usually directly related to the (rate of) inbreeding within the breed, and expressed as heterozygosity, effective population size, effective number of alleles per locus or Wright's F-statistics, usually also calculated from marker allele frequencies.

Conservation efforts should be as efficient as possible, securing a maximum amount of genetic diversity with the available resources. To this end, breeds at risk need to be evaluated for the amount of genetic diversity. The evaluation is very much dependent on the rationale for conservation (Ruane, 1999) and may require balancing diversity within and between populations.

There are two frameworks to quantify genetic diversity in a group of populations: the Weitzman diversity and the Core set diversity (section 5.3). The two frameworks rely on genetic distances or genetic similarities calculated from neutral marker allele frequencies. There are also other methods, such as cluster analysis (briefly discussed in chapter 4), multivariate analysis and principle component analysis.

1.2. Practical considerations in genetic diversity studies

In estimating genetic distance or kinship from marker data the sampling process is important. The relevant criteria involve the number of loci, including the number of alleles per locus, and the number of individuals. The individuals should be randomly drawn, to reflect the composition of the population. Generally, 25 sampled animals (N) are taken to be minimal (FAO, 1998). This results in 2N = 50 drawings of alleles per locus and gives a reliable estimate of allele frequencies, with low standard errors of genetic distance estimates. In very small populations it is worthwhile to sample the whole population. Then, the true gene frequencies are known. Often we will have different sample sizes between populations and can correct estimates for unequal sample sizes (Nei, 1987).

The genetic diversity measures try to detect kinship or similarity of alleles. Alleles of a locus in different members may be *identical by descent* due to inbreeding. However, alleles can also be indistinguishable from one another without descending from the same individual and they are said to be *alike in state*. For a correct estimate of genetic diversity it is important that the probability of two alleles being alike in state is minimised or we have a good understanding about the size of probability. This is achieved by using loci with as much polymorphism as possible. In the set of microsatellite markers proposed by FAO, the rule of thumb is at least 4 different alleles. The markers should follow simple Mendelian inheritance (Bretting and Widerlechner, 1995), therefore sex linked loci should be avoided or used with caution.

It should always be remembered that uniqueness of a breed is not determined by genetic considerations alone. Other considerations will be described later (chapter 6).

2. Genetic distances, F-statistics and kinships

The methods outlined produce statistics based on the (non-)similarity in allele frequencies between populations and give different interpretations for genetic diversity and conservation priorities. Genetic distances are used extensively in genetic diversity studies, especially for constructing the phylogeny of the populations. Wright's F-statistics are the standard instruments to investigate population sub-division and to partition genetic variation into between and within population components and may be used as indicators for the relative importance of breeds within a species.

More recently measures that estimate (mean) kinships between populations and/or between individuals within populations have been developed to specifically address questions in conservation genetics.

2.1. Genetic distances

Genetic distances are calculated from the squared differences in allele frequencies in two populations. Genetic distances have mathematical properties and biological significance. Mathematically, a function must have some properties to be a distance. First, the distance between a population X and itself must be zero, or d(X,X) = 0. Second, the distance between two populations X and Y must be symmetrical, or d(X,Y) = d(Y,X). If a distance satisfies these conditions, it is called a semi-metric distance. If between populations X, Y and Z a distance also satisfies the triangular inequality d(X,Y) ≤ d(X,Z) + d(Y,Z), it is called a metric distance (Katz, 1986).

The natural distance between two vectors, **x** and **y**, in a k-dimensional space (where k in this setting usually refers to the number of loci) is the Euclidean distance:

$$d(X,Y) = \sqrt{\sum_{i=1}^{k} (x_i - y_i)^2}$$

It can be shown that this distance satisfies the three mathematical properties. This was first used by Rogers (1972) genetic distance (D_{Rogers}). Note that its square satisfies the two first conditions but not the triangular inequality. Hence, distance measures based on the squared Euclidean distance (Nei's minimum distance, Reynolds distance, etc.) do not satisfy the triangular inequality, neither does Nei's standard genetic distance.

The biological interpretation of genetic distances is dependent on the divergence model assumed for population differentiation which has four forces: random drift, mutation, selection and migration. Since genetic distances were originally designed with species in mind, the respective models assume independent evolution for each population. After speciation (when two populations become two distinct species) there is by definition no migration between populations. Hence genetic distances ignore migration, which should be alleviated for farm animal populations. However, if migration does occur, the distances estimated are deflated.

The genetic distance between two farm animal populations is determined exclusively by random drift. Under a pure drift model, the inbreeding coefficient F is after t generations for an effective population N_e

$$F_t = 1 - \left(1 - \frac{1}{2N_e}\right)^t$$

For Nei's minimum distance (Nei 1973) - assuming p_{0m} is the founder population frequency of allele m - the expectation is:

$$E(D_M) = E\left(\frac{1}{2}\sum_m (p_{Im} - p_{Jm})^2\right) = \frac{1}{2}(F_I + F_J)\left(1 - \sum_m p_{0m}^2\right)$$

Under drift, the expectation of the usual genetic distances (like Nei's standard distance) between populations I and J is therefore $F_I + F_J$, the sum of the coefficients of within population inbreeding since population fission. The preferred distance should therefore have an expected value equal or proportional to $F_I + F_J$. Reynolds (1983) introduced a measure of genetic distance which is Nei's minimum distance normalised by an estimate of heterozygosity in the founder population $(1-\Sigma_m[p_{Im} \times p_{Jm}])$, effectively removing this part from the former equation. The expected value of Reynolds distance is therefore equal to $(F_I + F_J)/2$.

Under pure random drift, distance estimates do not reflect the exact phylogeny of the populations, as apart from the number of generations, it is influenced by effective population size. However, since the main interest is in conservation value of breeds rather than in phylogeny, this type of distance is useful for closely related populations, like breeds in Europe.

Phylogenetic trees are graphical representations or mappings of the matrix of the distances between populations. As we learnt, the trees are not phylogenetic, since differences in effective population size and migration between breeds may distort the picture. There are different methods of drawing distance matrix trees. Nei *et al.* (1983) discusses and compares the different methods. The most widely known methods are Neighbour-Joining (NJ) and UPGMA (Takezaki and Nei, 1996), which generally give good results.

UPGMA is somewhat simpler and easier to understand (see Box 5.2 'Constructing trees'). However, UPGMA assumes equal rates of evolution for all populations. Evolution rate is among others governed by the effective population size. Thus the rate of evolution will differ from one breed to the next, depending on their effective size. NJ takes varying rates of evolution into account and is therefore more appropriate for breeds. NJ also gives higher bootstrap values in most cases. Bootstrapping is a technique to assess the reliability of estimates via resampling data (Weir, 1990) and gives the possibility to draw multiple trees and estimate reliabilities for the different nodes in the tree. There are a number of software options to construct trees from distance data. Most of these programmes use genotype data and apply the desired distance and bootstrap options.

Box 5.1. Genetic distances.

For notation convenience x_i and y_i are frequencies of the ith allele respectively drawn in population X and Y. For simplification reasons, distance formulae are given for one locus. To extend those expressions for several loci, one has to sum over loci and divide by the number of loci where summations over alleles appear in the expressions.

Nei standard Genetic distance (D):

$$D = -\ln\left(\frac{\sum_i x_i y_i}{\sqrt{\sum_i x_i^2 \sum_i y_i^2}}\right)$$

The chord distance of Cavalli-Sforza (D_c):

$$D_c = (2/\pi)\sqrt{2\left(1-\sum_i \sqrt{x_i y_i}\right)}$$

Nei's distance (D_A):

$$D_A = 1-\sum_i \sqrt{x_i y_i}$$

Nei's minimum distance (D_m):

$$D_m = \frac{1}{2}\sum (x_i - y_i)^2$$

Reynolds distance ($D_{Reynolds}$):

$$D_{Reynalds} = \frac{1}{2}\frac{\sum (x_i - y_i)^2}{1-\sum x_i y_i}$$

Interpretation of phylogenetic trees should be done with the utmost care. A major assumption in phylogenetic trees is isolation after population fission. Generally this is an assumption that will not hold in the case of livestock populations. There might be gene flow between populations and even between clusters of populations, giving back some common features to populations once having been separated in fission. This will

Box 5.2. Constructing trees.

Example of the construction of a tree using UPGMA. Although NJ, in general, gives better results, UPGMA is more suitable for illustrating the process.
Suppose we have four breeds A, B, C and D. The distances between them are given below.

	B	**C**	**D**
A	.400	.300	.500
B		.200	.100
C			.300

We start with the pair of breeds that is closest to one another. The closest pair is (B, D). Therefore we next calculate the distances between the cluster (B,D) and A (or C) as the average of A's distance to B and D. Therefore distance ((B, D), A) = ½(.400 +.500)=.450

	(B,D)	**C**
A	.450	.300
(B,D)		.250

Again we find the smallest distance (between (B, D) and C) and recalculate the distance between this cluster and A (giving (B, D, C), A =.375). The resulting unrooted tree is drawn below. The branches are drawn in such a way the lengths of the branches between two breeds sum up to the distance given above.

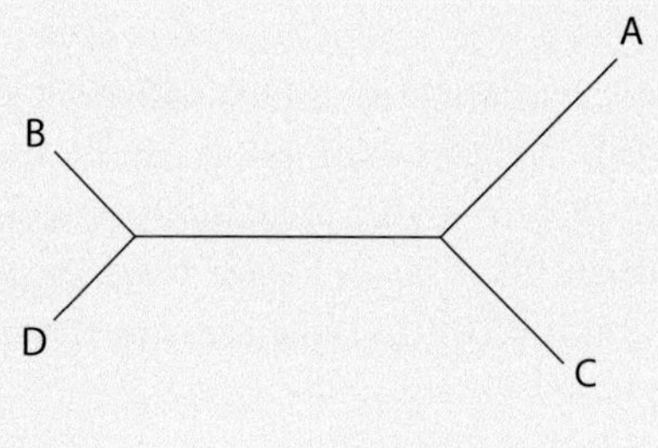

have an effect on the distribution of genetic variation between and within populations, an effect that may not be concluded from the tree presentation. On the other hand, migration between populations is reflected as smaller distances. While this leads to errors in the estimated phylogeny, it is nonetheless a good indicator of the closeness of populations. It does not, however, correctly place the shared ancestor. Trees from drift-based genetic distances should not be interpreted as actual phylogeny.

2.2. F-statistics

Random drift is seen between populations as differences in allele frequencies and loss of different alleles. Therefore the effects of drift within and between populations are in opposite direction: Within populations genetic diversity will be lost, while between populations genetic differentiation will increase. This phenomenon is formulated in the well-known expression of Wright's F-statistics (Wright, 1969):

$$(1-F_{IT}) = (1-F_{IS})(1-F_{ST})$$

where F_{IT} is the inbreeding coefficient of an individual relative to the whole set of populations, F_{IS} is the inbreeding coefficient of an individual relative to the sub-populations it belongs to and F_{ST} is the mean inbreeding coefficient of sub-population relative to the entire population. Note that this expression assumes that a set of populations under study are descendant (or sub-populations) of one population.

The inbreeding coefficient F in a population can be computed from the differences between expected and observed homozygosity:

$$HOM_{exp} = \sum_i p_i^2$$

and

$$HOM_{obs} = F + (1-F)Hom_{\exp}$$

Since the decrease in genetic variance in a population is proportional to $(1 - F)$ (Falconer and Mackay, 1996), Wrights expression can be used to partition the total genetic variance of populations into a within population component $(1 - F_{IS})$ and a between population component $(1 - F_{ST})$. Proportions of genetic variation between and within populations are obtained by dividing these by the total variation component estimated by $(1 - F_{IT})$.

Within a population, F_{IS} is usually estimated from the excess of homozygotes (or reversely, the deficit of heterozygotes):

$$F_{IS} = \frac{\overline{Het}_S - Het_{obs}}{\overline{Het}_S}$$

where $\overline{Het}_S$ is the mean of expected heterozygosities in populations. Multiple estimates over loci are averaged to obtain a mean estimate. Wright's F_{ST} statistic is a popular measure of population subdivision and differentiation. The estimation of F_{ST} is deduced from the relationship:

$$Het_{ST} = 1 - F_{ST} = \frac{\overline{Het}_S}{Het_T}$$

The expected heterozygosity in the total population Het_T is calculated from the allele frequencies in the total population, p_{Ti}, i.e. $Het_T = 1\text{-}\Sigma p_{Ti}^2$ (Nagylaki, 1998).

There are a number of variations on the way F_{ST} is calculated. Weir and Cockerham (1984) and Robertson and Hill (1984) give different estimators of F_{ST} from allele frequencies. The estimator of Robertson and Hill gives extra weight to rare alleles for conservation purposes. However, the variance of the estimator is greater and both estimators agree only when alleles have equal frequencies.

Nagylaki (1998) argues that F_{ST} will only be an appropriate measure of divergence of populations if the genetic diversity is low. For example: when we have N sub-populations of equal size, which do not share alleles at all, then

$$F_{ST} = \frac{(N-1)(1-Het_S)}{N-(1-Het_S)}$$

Except when the populations are fully inbred ($Het_S = 0$), F_{ST} will always be smaller than 1, even though the sub-populations are fully differentiated. Moreover, when we have K sub-populations fixed for a locus with L < K alleles, the average heterozygosity within populations will be 0 and hence $F_{ST} = 1$, indicating complete differentiation between sub-populations. However, L < K means that complete differentiation is only possible for L sub-populations. When K > L complete differentiation cannot be obtained and the F_{ST} value of 1 is misleading.

2.3. Genetic similarities and kinships

While genetic distances and F-statistics are interested in differences between populations or individuals, the measure on similarity is interested in resemblances between them. The genetic similarity measures the degree of relatedness, which is a complement to distance, hence 1 minus the similarity between two individuals or populations is roughly equal to the distance between them. Relatedness is usually expressed as coefficient of kinship.

Relating coefficients of kinship to genetic diversity is straightforward. Over t generations, the loss in heterozygosity is directly related to the inbreeding coefficient:

$$Het_t/Het_0 = 1 - F$$

where Het_t is heterozygosity in generation t and Het_0 in the founder generation, and F is the inbreeding coefficient relative to the founder generation. Kinship, also called coancestry (f), is used to calculate the inbreeding coefficient and $F_X = f_{PQ}$, where f_{PQ} is the coancestry of the parents P and Q of individual X. Twice the kinship, the coefficient of additive relationship is used to calculate the additive genetic variance σ^2_A. Because σ^2_A is proportional to heterozygosity, over t generations we have (Falconer and Mackay, 1996; Gilligan *et al.*, 2005):

$$\sigma^2_{A,t}/\sigma^2_{A,0} = 1 - F$$

There are many different estimators for relatedness. There are coancestry based estimators (Toro *et al.*, 2002; Eding and Meuwissen, 2001; Oliehoek *et al.*, 2006) and two- and four-gene identity coefficient estimators (Lynch and Ritland, 1999; Wang, 2002). Coancestry based estimators are more general in nature and perform well in a wide variety of population classes, while two- and four-gene identity coefficients show a substantial loss of efficiency in non-random mating populations (Oliehoek *et al.*, 2006). For these reasons we focus on coancestry based estimators of kinship.

2.3.1. Genetic similarities

The most basic measure of relatedness is the genetic similarity. Genetic similarities are also known as allele sharing (cf. Lynch, 1988). Basically any pair of individuals within or between populations is scored for common alleles for a number of loci. The total score is then averaged over loci to obtain the mean similarity between individuals. Further averaging over pairs of individuals gives the mean similarity between or within populations.

There are two main methods to score genetic similarities with co-dominant polymorphic markers: the genic similarity (Lynch, 1988) and the Malécot similarity (Eding and Meuwissen, 2001). The difference between these two can be found in scoring of similar genotypes (Table 5.1). While the genic similarity is the number of alleles shared by two individuals out of the total number of alleles (e.g. 4 in diploid organisms), the Malécot similarity is the probability that an allele randomly drawn from one individual is the same as an allele randomly drawn from the other individual (Malécot, 1948). The latter is derived from the definition of the coefficient of kinship. Hence, if we assume that alleles can only be identical by descent (that is: all similar alleles are copies from one and the same ancestral allele), the mean Malécot similarity (calculated over multiple loci) is expected to be equal to the coefficient of kinship.

Table 5.1. Scoring of similar genotypes in the genic and Malecot genetic similarity and probabilities in populations at Hardy-Weinberg equilibrium.

genotype	genic	Malecot	Probability
AA-AA	1	1	$\sum_{m=1}^{M} p_{I,m}^2 p_{J,m}^2$
AA-AB	½	½	$2\sum_{m=1}^{M} p_{I,m}^2 p_{J,m}(1-p_{J,m})$
AB-AA	½	½	$2\sum_{m=1}^{M} p_{I,m}(1-p_{I,m})p_{J,m}^2$
AB-AB	1	½	$2\sum_{m=1}^{M}\sum_{n\neq m}^{M} p_{I,m}p_{J,m}p_{I,n}p_{J,n}$
AB-AC	½	¼	$2\sum_{m=1}^{M}\sum_{n\neq m}^{M} p_{I,m}p_{J,m}p_{I,n}(1-p_{J,m}-p_{J,n}) + 2\sum_{m=1}^{M}\sum_{n\neq m}^{M} p_{I,m}p_{J,m}(1-p_{I,m}-p_{I,n})p_{J,n}$

In formula form the Malécot similarity for two individuals with genotypes a/b and c/d at locus k is written as:

$$S_{xy,k} = \frac{1}{4}[I_{ac} + I_{ad} + I_{bc} + I_{bd}]$$

where I_{xy} is an indicator variable that is 1 when allele x and y are similar, otherwise I_{xy} is zero. Thus the similarity can have values 0, ¼, ½ and 1.

The Malécot similarity is advantageous over other similarity measures because it can be calculated directly from allele frequencies (Eding and Meuwissen, 2001). For a locus with M alleles the similarity between populations I and J is:

$$S_{IJ} = \sum_{m=1}^{M} p_{I,m} p_{J,m}$$

The expression for population similarity can be found in different guises in different genetic diversity measures, showing that there is a strong connection between genetic diversity and kinship (Box 5.3).

2.3.2. Correction for alleles alike in state

Technically indistinguishable alleles are either *identical by descent* (IBD, two alleles are copies of the same ancestral allele due to kinship, or *alike in state* (AIS, two alleles are indistinguishable from one another, but are not IBD). The probability of alleles AIS is indicated with the symbol s.

The mean expected similarity S_{ijl} between two individuals i and j for a locus l is a function of both the kinship between i and j (f_{ij}) and s_l at this locus (Lynch, 1988):

$$S_{ijl} = f_{ij} + (1 - f_{ij})s_l = (1 - s_l)f_{ij} + s_l$$

Rearrangement of the equation, substituting expected similarities with observed similarities and averaging over loci gives an estimator of f_{ij} for L loci:

$$\hat{f}_{ij} = \frac{1}{L}\sum_{l=1}^{L}\frac{S_{ijl} - s_l}{1 - s_l}$$

Thus, to estimate kinships between individuals (or populations) some value of s_l must be assumed or estimated.

If we assume a model with an infinite number of alleles in the founder population, s_l equals zero and f_{ij} is expected to be equal to S_{ijl} for all loci. However, when s_l is non-zero, a founder population, in which all individuals are assumed unrelated ($f = 0$), will have $S_l = 0 + (1 - 0)s = s_l$. Hence definition of s_l implicitly defines the founder population.

2.3.3. Molecular coancestry

(Toro *et al.*, 2003) estimates coefficient of kinships under the assumption that $s_l = 0$ for all loci. Between and within populations the molecular coancestry is the average over loci of the Malécot similarity. Between individuals, this similarity can be expressed as:

$$f_{M,ij} = \overline{S_{ij}} = \frac{1}{L}\sum_{l=1}^{L}\frac{1}{4}[I_{ac} + I_{ad} + I_{bc} + I_{bd}]_l$$

Box 5.3. Connections between genetic distances, similarity and kinship coefficients.

Under pure random drift the Malécot similarity offers a good opportunity to demonstrate the close relations between genetic distances and kinship coefficients. The similarity is calculated from allele frequencies as:

$$S_{IJ} = \sum_{m=1}^{M} p_{I,m} p_{J,m}$$

Similarities expressed in terms of allele frequencies appear in a number of genetic distance measures. Given that $S_{IJ} = s + (1 - s)f_{IJ}$ (where s is the probability of alleles AIS), the distance using the expression is a function of kinship coefficient and the probability of alleles AIS. For instance, if we ignore the probability of alleles AIS (assume $s_l = 0$), Nei's standard distance reduces

$$D = -\ln\left(\frac{f_{IJ}}{\sqrt{f_I f_J}}\right)$$

Another recurring expression in genetic distances is the sum of squared differences between allele frequencies:

$$\sum (p_{I,m} - p_{J,m})^2$$

If we write this out we obtain:

$$\sum (p_{I,m} - p_{J,m})^2 = \sum_{m=1}^{M} (p_{I,m})^2 + \sum_{m=1}^{M} (p_{I,m})^2 - 2\sum_{m=1}^{M} p_{I,m} p_{J,m}$$

$$\Rightarrow$$

$$\sum (p_{I,m} - p_{J,m})^2 = S_I + S_J - 2S_{IJ}$$

The squared differences between allele frequencies can thus be written in terms of mean kinships as:

$$\sum (p_{I,m} - p_{J,m})^2 = (1 - s)[f_I + f_J - 2f_{IJ}]$$

Thus genetic distances and population kinships are closely related. The differences between distance measures result from a different scaling of the kinships. Unlike distances, kinships are directly related to standard genetic theory and thus offer more direct interpretation of estimates.

Although performing better than non-Malécot genetic similarities, both in accuracy and in correlation with pedigree data, it does not take into account that the variance in similarity between loci is caused by the variance in s_l.

2.3.4. Current homozygosity

Li *et al.* (1993) proposed an estimator in which s_l was set to equal the current homozygosity within the population. This assumption places the founder population in the same era as the current population, since $S_{founder} = s$, and the mean estimated kinship between individuals is expected to equal -1/2N. While closely related individuals will show a positive kinship, more distantly related pairs will have negative coancestry. This is not a desired outcome, since kinship is equal to the probability of alleles IBD and hence must be equal or greater then zero by definition.

2.3.5. Minimal mean similarity

A key property of mean kinships (and genetic similarities) between populations is the 'stationary property'. After fission of two populations with subsequent complete isolation, the mean kinship (and the mean similarity) between populations will settle within a few generations on a value equal to the mean kinship (or similarity) within the population that existed just prior to fission. In a study on multiple populations the pair of populations with the lowest mean genetic similarity (averaged over loci) is expected to reflect the founder population at the time when the very first fission of the founder population occurred. Hence s_l can be set to S_{ijl} of the pair with the lowest mean similarity, setting the mean kinship between the selected pair to zero. Subsequently, all other mean kinships can be estimated using the estimator described before. Using this method will yield estimates that are expected to be positive and more accurate then f_M, since variation between loci for s_l is accounted for in a conservative way.

A further refinement can be made by applying weights to observed similarities. This ensures that more informative markers (more polymorphic and more uniformly distributed allele frequencies) will influence the estimate more than less informative ones.

2.3.6. Log-linear models

A method that is free of *a priori* assumption about the probability of alleles AIS is the use of linear regression. The expression relating S_{ijl} to f_{ij} and s_l is log-transformed to

$$\ln(1 - S_{ij,l}) = \ln(1 - f_{ij}) + \ln(1 - s_l) + error_{ij,l} \quad \Leftrightarrow \quad y_{ij,l} = a_{ij} + b_l + error_{ij,l}$$

Thus, the kinship between a pair of populations or individuals is expected to be constant over all loci, while the probability of alleles AIS (s_l) is expected to be equal for all pairs of populations within a locus.

This can be extended to N breeds and L marker loci using weighted loglinear model (WLM) (see Eding and Meuwissen, 2003 for details). All observed similarities are given weights to account for information content. The weights are the (expected) variance of the similarities.

The WLM method has a few drawbacks. First, the estimation expands quadratically with the number of populations or individuals. Solving the linear regression equations requires large amounts of CPU time and memory. This makes the WLM not particularly suited for analysis on the level of individuals. Second, because of the log-transformation of the data, similarities of $S_{ij} = 1$ lead to $\ln(0) = -\infty$. This can be solved by changing $S_{ij} = 1$ to $S_{ij} = 0.9999$, but at the cost of accuracy of estimation. Particularly when a population is highly homozygous and fixation occurs with a high frequency, estimates of kinship become unreliable (Eding and Meuwissen, 2003; Oliehoek *et al.*, 2006).

2.3.7. Weighted Equal Drift Similarity

Much simpler then the Weighted Loglinear Model is the Weighted Equal Drift Similarity (*WEDS*; Oliehoek *et al.*, 2006). The correction for alleles AIS in the WEDS method is a compromise between assumed and estimated s_l that works well enough to be applicable to large sets of individuals. The correlation between estimated kinships and actual kinships (calculated from pedigrees) is relatively high, independent of population structure. This makes the method very suitable for analysis of individuals within a population.

The estimation of Marker Estimated Kinships with *WEDS* is a three step procedure:

1. **Calculate s_l:** This step starts with setting to zero the s_l for the locus with the lowest expected similarity across individuals, S_{min}. Thus the similarity of the most polymorphic locus is assumed to be equal to the kinship coefficients. All other similarities are then adjusted through:

$$\hat{s}_l = \frac{S_l - S_{min}}{1 - S_{min}}$$

2. **Calculate weights w_l:** As in the Weighted Loglinear Model described above, weights are used to account for varying degrees of informativeness. The weights are the inverse of the expected variance of the observed similarity.
3. **Estimate kinships:** Lastly the kinships are estimated as the weighed average of per locus similarities corrected for alleles AIS, using the weights and probability of alleles AIS calculated in the previous steps.

The WEDS method of estimation does not require large numbers of equations to be solved simultaneously, reducing the computer resources needed to analyse large data sets.

2.3.8. Bootstrapping procedures in kinship coefficient estimation

Bootstrapping can be used in the estimation procedures to obtain more accurate kinship estimates. The simplest way is a bootstrap over loci (randomly drawing, with replacement, loci from the panel of used markers), but Bennewitz and Meuwissen (2005) reported more accurate similarities scores, if a parallel bootstrap was performed within populations over individuals.

2.3.9. Visualisation

Kinships between and within populations can be visualised with a quasi-phylogenetic tree or a contour plot (Eding *et al.*, 2002; Mateus *et al.*, 2004).

A *phylogenetic tree* constructed with standard software like PHYLIP (Felsenstein, 1996) starts from converting the kinship matrix to a kinship-distance matrix (Eding *et al.*, 2002):

$$d(i,j) = \hat{f}_{ii} + \hat{f}_{jj} - 2\hat{f}_{ij}$$

where f_{ii} (f_{jj}) is the within population kinship estimate for population i(j) and f_{ij} is the estimate of the kinship between populations i and j. Note that this distance is equal to twice Nei's minimum distance corrected for the probability of alleles AIS. The reservations about interpreting trees on genetic distances (paragraph 2.1) apply here as well.

The clustering order obtained by constructing the phylogenetic tree can be used to re-arrange the matrix containing kinship estimates. When a *contour plot* of the re-arranged matrix is made by shading according to estimated kinship values, cross-relatedness between clusters and patterns of gene flow can be better appreciated (Figure 5.1).

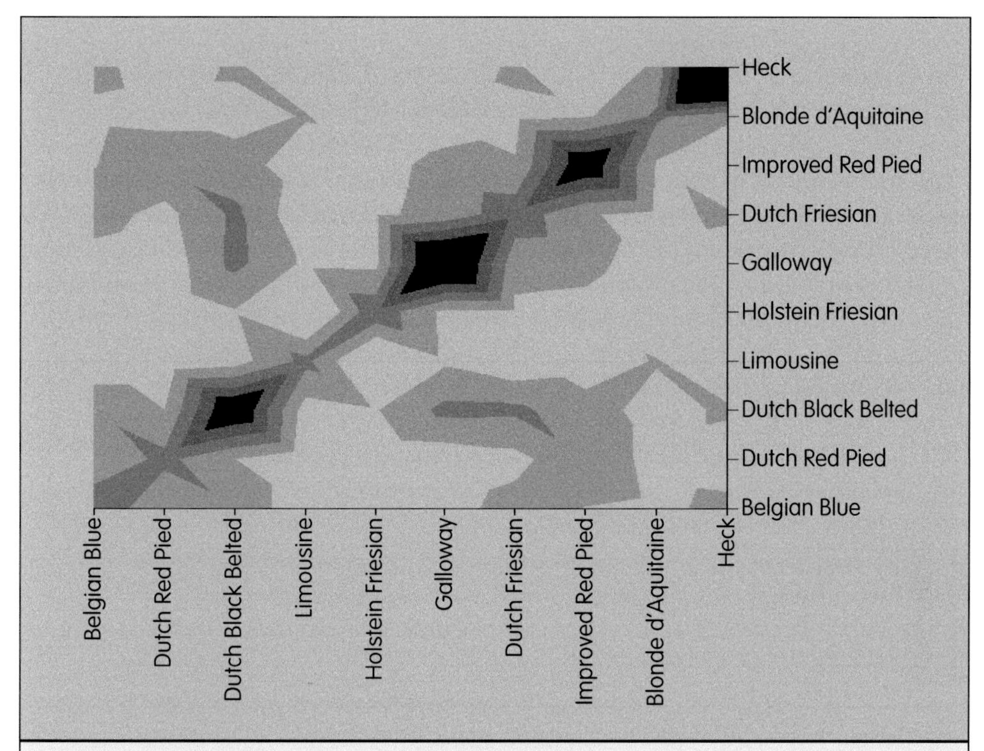

Figure 5.1. Contour plot of the estimated kinship matrix of a small data set of Dutch cattle populations. Darker shading indicates higher kinship, e.g. the elevated kinship of the Dutch Black Belted with Dutch Friesian and Galloway (data taken from Bennewitz and Meuwissen, 2005).

3. Weitzman and Core set diversities

In deciding which populations or individuals are the most important contributors to genetic diversity, we need quantitative assessments, whilst the methodology we have discussed so far provided only qualitative information. In this section we will present two quantitative methods: The Weitzman diversity method and the Core set diversity method.

3.1. Weitzman diversity

The Weitzman method of determining (genetic) diversity is a recursive algorithm to calculate the total diversity in a set and to determine the relative importance of elements (breeds, individuals) to total diversity, based on genetic distances. The method adheres to the criteria for a good diversity measure set by Weitzman (the Weitzman criteria, see

Box 5.4). However, due to the nature of genetic distances, it nominally accounts only for between breed diversity and neglects within breed diversity.

For a set *S* of *N* breeds with pair wise distance $d(i,j)$, between breeds *i* and *j*, a diversity metric $D(S)$ can be computed from an $N \times N$ distance matrix with a recursive algorithm suggested by Weitzman (1992). The methodology also yields a tree with maximum likelihood properties. The contribution of an element is proportional to the reduction

Box 5.4. Weitzman criteria for proper diversity measures.

Weitzman (1992) defined four criteria for a proper measure of diversity:

- *Criterion 1: Continuity in species.* The total amount of diversity in a set of populations should not increase when a population is removed from the set.
- *Criterion 2: The twin property.* The addition of an element identical to an element already in the set should not change the diversity content in a set of populations.
- *Criterion 3: Continuity in distance.* A small change in distance measures should not result in large changes in the diversity measure.
- *Criterion 4: Monotonicity in distance.* The diversity contained in a set of populations should increase if the distance between these populations increases.

Both the Weitzman method (not to be confused with the Weitzman criteria) and the Core set method satisfy these criteria despite the difference in the source information. Since relatedness measures are essentially measures of variance, it is possible that the genetic diversity in terms of mean kinships or relatedness increases when a population is removed from the set (Thaon d'Arnoldi *et al.*, 1998). However, when the contribution of each population is optimised (see paragraph 3.3.), the mean kinship is at a minimum and the genetic diversity is at a maximum. Removal of a breed that has a non-zero contribution will therefore decrease genetic diversity (criterion 1). If a population is identical to another one, its contribution is zero and it can be excluded (criterion 2). The measure of genetic diversity is a continuous function of the (estimated) mean kinships between and within breeds and the measure changes only slightly, when kinships or distances change slightly (criterion 3).
With regards to criterion 4, an increase in genetic distance in a pure drift model can be caused by: (1) a decrease in the kinship between breeds, and (2) an increase in the within breed kinships (i.e. continued inbreeding within a population). In the latter situation, criterion 4 does not hold, since continued inbreeding reduces genetic diversity, even if the genetic distance increases. Criterion 4 can be rewritten in terms of kinships between populations as follows: the diversity contained in a pair of populations should increase if the kinships between or within these populations decrease.

in tree length caused by its removal from the group. The total diversity of set S is defined recursively as:

$$D_W(S) = \max_{i \notin S}\left[D_W(S_{i \notin S}) + d(i, S_{i \notin S})\right]$$

where is the distance between i and its closest member in S. Although this method was first conceived for species (hence the reliance on genetic distances) it has been applied to livestock breed diversity by Thaon d'Arnoldi *et al.* (1998) and Reist-Marti *et al.* (2003). Attempts to incorporate within breed diversity into the Weitzman approach sought to remedy the limitation to between breed diversity only (Garcia *et al.*, 2005). When Weitzman diversity is based on genetic distances, it has a tendency to put emphasis on populations that have drifted furthest and have undergone inbreeding (see paragraph 5).

3.2. Core set diversity

Core set diversity is based on measures of coancestry or kinship (Eding *et al.* 2002; Bennewitz and Meuwissen, 2005; Oliehoek *et al.*, 2006). The concept of a core set first appeared in plant conservation genetics and was defined as the smallest set of lines or strains of a plant species that still encompasses the genetic diversity in the species (Frankel and Brown, 1984). The aim is the elimination of genetic overlap between lines in the core set. The genetic overlap or genetic similarity between individuals and populations is described by the coefficient of kinship. Hence eliminating genetic overlap is equal to minimising kinship in a set of breeds by adjusting the contribution of each population or individual to the core set. We can maximise genetic diversity and find the relative importance of populations or individuals in conserving the genetic diversity.

As an example to illustrate the principle of core sets, consider three populations, where populations 2 and 3 are identical, while population 1 is unrelated to both 2 and 3. The kinship matrix is:

$$\mathbf{K} = \begin{bmatrix} 1 & 0 & 0 \\ 0 & 1 & 1 \\ 0 & 1 & 1 \end{bmatrix}$$

The mean kinship in $\mathbf{K}$ is $5/9 \approx 0.56$ (5 ones over 9 elements). Removal of population 3 from $\mathbf{K}$ leads to

$$\mathbf{K}^* = \begin{bmatrix} 1 & 0 \\ 0 & 1 \end{bmatrix}$$

and the mean kinship has decreased to 2/4 = 0.50, which implies an increase in genetic diversity. This is in violation of the Weitzman criteria (Box 5.4): the removal of a population should have either a negative or zero effect on the total diversity.

The decrease in mean kinship, as a result of the removal of population 3 from the set, occurred because populations 3 and 2 are identical. There is one population that contributes twice to the mean kinship and is actually over-represented. The over-representation is avoided by basing the diversity on the mean kinship. The core set is a mixture of populations so that "genetic overlap" is minimised. The genetic overlap is eliminated by removing population 3 (or equivalently, population 2)(Eding *et al.*, 2002). Alternatively the contributions of populations 2 and 3 can be set to half the contribution of population 1.

3.3. MVO and MVT core sets

The set of contributions of each population that minimises the mean kinship in S (i.e. contributions to the core set) can be calculated in different ways.

3.3.1. Maximum variance in offspring

The most straightforward measure is the maximum genetic variance in the offspring, or MVO core set (Eding *et al.*, 2002; Caballero and Toro, 2002). In this method the genetic variance of a non-specified quantitative trait is maximised within a hypothetical population bred from the populations or individuals that make up the core set. Variation between and within entities receive equal weight. The optimised contributions to a MVO core set are calculated through:

$$\mathbf{c}_{\mathbf{mvo}} = \frac{\mathbf{1'K}^{-1}}{\mathbf{1'K}^{-1}\mathbf{1}}$$

Where **1** denotes a vector of ones of size equal to the number of populations; $\mathbf{K}^{-1}$ is the inverse of the estimated kinship matrix. The genetic diversity in a set of populations S is defined as 1 – the mean kinship in the set:

$$D_{MVO}(S) = 1 - \bar{f}_S = 1 - \mathbf{c}_{\mathbf{mvo}}'\mathbf{K}\mathbf{c}_{\mathbf{mvo}}$$

The elements in $\mathbf{c}_{\mathbf{mvo}}$ sum up to 1.

Since the founder population is assumed to have a mean f of zero and $D_{MVO}(founder) = 1$, $D_{MVO}(S)$ is equal to the fraction of genetic diversity in the founder population that survives in the S populations.

The MVO core set places emphasis on populations with both low within and between population kinships. While this conserves a maximum of variation, it also means that breeds under threat of extinction, by definition having small (effective) population size, will be at a disadvantage. Such populations will contribute small amounts of genetic diversity as a result of increased genetic drift and inbreeding and of small within population variance.

Box 5.5. Within and between population variation.

The total genetic variance, or total gene diversity GD_T is the sum of the genetic diversity between populations GD_B and within populations GD_W:

$$GD_T = GD_B + GD_W$$

Scenarios of conservation can be envisioned that require putting more emphasis on either one of these. In formula form this can be written as:

$$GD_T = GD_B + \lambda GD_W$$

Where λ is a factor determining the importance of within population genetic diversity relative to the between population diversity.

Both MVO and MVT core sets are derived from standard quantitative genetic theory but they implicitly put different emphasis on the between and within components. The MVO core set has equal weights or $\lambda = 1$. The MVT core set can be shown to put more weight on GD_B: $\lambda = ½$ (Toro *et al.*, 2006).

It can be argued that variation between populations is more important, because the important traits are expressed by genes that will likely be fixed within populations. Pyasatian and Kinghorn (2003) suggested to put five times more emphasis on between population diversity ($\lambda = 0.2$). The difference in weight relates to the difference in speed with which genetic change can be realised across populations or within a single mixed population. Toro *et al.* (2006) show how different values of λ change priorities of populations for conservation.

3.3.2. Maximum Variance Total

The maximum variance total or MVT core set (Bennewitz and Meuwissen, 2005) does not maximise the genetic variance within a hypothetical, but maximises the total genetic variance within and between the populations. The optimised contributions to an MVT core set are calculated through:

$$\mathbf{c}_{\mathbf{mvt}} = \frac{1}{4}\left(\mathbf{K}^{-1}\mathbf{D} - \frac{\mathbf{1'K}^{-1}\mathbf{D} - 4}{\mathbf{1'K}^{-1}\mathbf{1}} \cdot \mathbf{K}^{-1}\mathbf{1}\right)$$

where $\mathbf{D} = \text{diag}(\mathbf{K})$ is a vector of size equal to the number of populations containing the diagonal elements of **K**. The genetic diversity in this set is calculated as:

$$D_{MVT}(S) = 1 + \mathbf{c'}_{\mathbf{mvt}}\,\mathbf{D} - 2\mathbf{c'}_{\mathbf{mvt}}\,\mathbf{K}\mathbf{c}_{\mathbf{mvt}}$$

As before, $D_{MVT}(S)$ is the total amount of genetic diversity relative to the genetic diversity in the founder population. D_{MVT} is expressed in terms of trait mean variance, and not of total genetic variance, and the MVT core set will sometimes produce diversity estimates larger then 1, indicating that the variance in the core set is more then the variance present in the (hypothetical) founder. This is due to evaluating the variance in separate populations. According to genetic theory, the additive genetic variance lost due to inbreeding within a population doubles the additive genetic variance gained between populations:

$$Var(G_{total}) = (1 - f)Var(G_{within}) + 2f\ Var(G_{total})$$

Therefore results of $D_{MVT} > 1$ are not contradictory to standard genetic theory.

The MVT core set places more emphasis on populations that have high within population kinship, but low between population kinships. It tends to value small populations that are distinct from the rest of *S*. MVT maximises the total variance, including the variance between populations, for quantitative traits. Higher contributions are (theoretically) given to breeds with more diverse phenotypes. Thus the variance is maximised in such a way that it would be easier for breeders to focus on a specific set of traits from the MVT core set.

4. Genetic distances, kinships and conservation decisions

Being proportional to the time span since divergence, genetic distances create the impression of increasing diversity between two populations, even when there is no

Box 5.6. Example of MVO and MVT core sets.

As an illustration to the differences between MVO and MVT we present the results from Dutch cattle populations. A contour plot of the kinship matrix of this small set was given in Figure 5.1. The kinship matrix was estimated using the Weighted Log-linear Model and bootstrapping over loci and individuals (Bennewitz and Meuwissen, 2005).

	MVO	MVT
Limousine	0.304	0.169
Holstein Friesian	0.229	0.209
Dutch Red Pied	0.115	0.028
Dutch Friesian	0.086	0.004
Heck	0.072	0.267
Galloway	0.051	0.139
Improved Red Pied	0.049	0.133
Blonde d'Aquitaine	0.039	0
Belgian Blue	0.032	0.002
Dutch Black Belted	0.021	0.049

The most striking difference is the difference in valuation of the Heck population. The Heck population in the Netherlands was started from a very limited number of founder animals. Within the population animals are highly related to each other. Nevertheless, the genetic background of the breed (mostly East and South European breeds; Felius, 1995) is reflected in a low mean kinship between this population and all others. The MVO gives the Heck population moderate priority, as it is inbred to a degree that exceeds that of all other populations. The MVT, on the other hand, puts high value on the Heck population being so distinct form the rest of the population.

change in the actual genetic diversity in terms of allelic diversity or coefficient of kinships. The mean kinship within a population can be written as:

$$f_i = f_{ij} + \Delta f_i$$

That is: the mean within population kinship is the sum of the mean kinship of the founder of f_{xy} and the increase in within population kinship since fission (Δf_x).

The total distance between a pair of populations *i* and *j* is determined by two distances: the distance between each population and the most recent common ancestor of *i* and *j* (i.e. the founder of the pair (*i*, *j*); Eding and Meuwissen, 2001):

$$d(i,j) = f_i + f_j - 2f_{ij} = (f_i - f_{ij}) + (f_j - f_{ij})$$

$$\Leftrightarrow$$

$$d(i,j) = \Delta f_i + \Delta f_j \approx \Delta F_i + \Delta F_j$$

Essentially the distance between *i* and *j* is determined by the increase in kinships (or the amount of inbreeding) since the founder of *i* and *j*. Given that f_{ij} remains unchanged after population fission, an increase in distance between *i* and *j* can only be caused by an increase in f_i and/or an increase in f_j. This means that in this case an increase in distance can only occur if the inbreeding coefficient in *i* and/or *j* increases. In other words, given that f_{ij} remains constant after population fission, an increase in distance is associated with a loss of (within population) genetic diversity.

A larger genetic distance is only related to a larger diversity if the within population kinships are equal. If within population kinships vary, a larger distance may lead to lower diversity, as the following example illustrates: Suppose there is a phylogenetic tree as given in Figure 5.2. In this figure the similarity scores are given within and between breeds. Nei's genetic distances between (A,B) (A,C) and (B,C) are given in the table. Mean kinship coefficients calculated from the similarities are also given.

If two populations were chosen for conservation based on these distances, the choice would be the pair (A,B) as they have the largest distance between them and seem the furthest apart. However, both the within and between population kinship is smaller (and consequently the conserved diversity larger), when the pair (A,C) or (B,C) is chosen for conservation instead of (A,B).

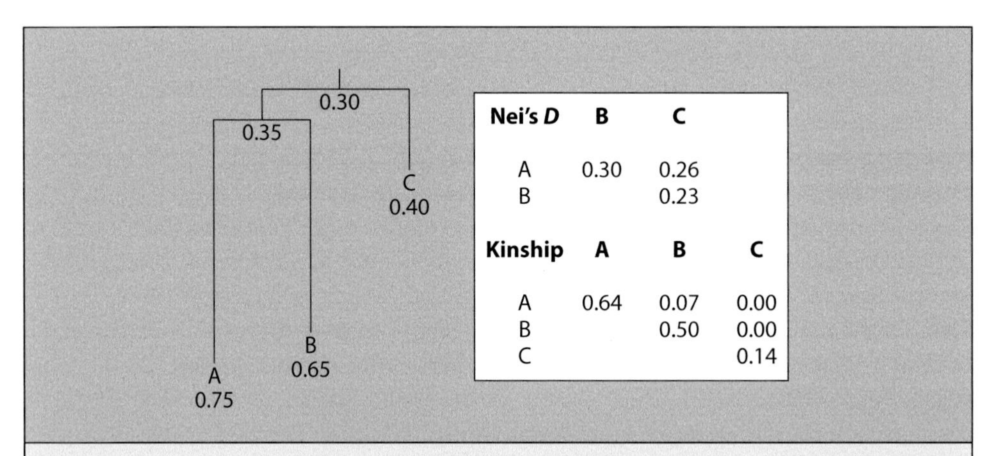

Nei's *D*	B	C
A	0.30	0.26
B		0.23

Kinship	A	B	C
A	0.64	0.07	0.00
B		0.50	0.00
C			0.14

Figure 5.2. Example of a phylogenetic tree between breeds A, B and C, and calculated Nei's genetic distances and mean kinship coeffcients.

Applying the Weitzman method results in the choice for populations A and B, with C as the link element in the diversity tree. This implies that the loss of population C has less consequences for the conservation of genetic diversity than the loss of any of the other element. Clearly, the loss of population C in the present example would yield the highest loss of diversity.

The relation between kinships and genetic distances noted above and elsewhere in this chapter leads to a fundamental difference between Weitzman diversity, which is based on genetic distances, and the Core set diversity, which is based on kinship coefficients. Weitzmans original criterion 4, the assertion that diversity should increase when the genetic distance increases, favours populations with extreme allele frequencies. On the other hand, the kinship based Core set diversity will decrease when extreme allele frequencies occur (i.e. due to high inbreeding in the population). Favouring populations with extreme frequencies implies that homozygote populations are (positively) valued. Kinship based diversity does not value homozygotes, but values the genetic variance in a random mating population that could be bred from the conserved set of populations. Conservation plans that maximise kinship based diversity will minimise the change in allele frequencies from the founder population and thus also minimise the loss of alleles.

Chevalet *et al.* (2006) carried out a simulation study comparing the ability to conserve alleles and heterozygosity by the Weitzman diversity and the Core set diversity approaches in a number of situations. They found the Core set approach to be more reliable than the Weitzman approach, precisely because the Weitzman diversity is based on genetic distances.

5. Concluding remarks

Over recent years the field of genetic diversity assessment has experienced a remarkable development. The definition of the Weitzman criteria was an important step in this development. It provides the framework for proper genetic diversity assessment. It holds for measures based on genetic distances (the Weitzman method) as well as for coancestry based methods (the Core set methods), although these criteria need proper formulation within the context of the measure used (Box 5.4).

Although genetic distances have been used for a long time in diversity studies, they have limited use in assessing genetic diversity. It was always assumed that a large distance equates to much diversity, but this is only the case if the within population genetic diversity would remain constant over time. However, the genetic diversity within a population decreases over time, depending on the effective population size. This process is not accounted for in genetic distances, since they are expressed between entities only. Moreover, the within population diversity is obscured in genetic distances, but plays a non-trivial role in determining the size and scaling of the distance between breeds (see Box 5.3). This also explains the sometimes counter-productive results that Weitzman diversity calculations produce (paragraph 4.). Finally, while scaling differs from one distance measure to the next, they all are closely related to kinship and inbreeding. However, none of them relates as directly to (population) genetic theory as do methods that measure kinship directly. Hence, kinship based methods of assessing genetic diversity, like Core set diversity, are preferable.

The choice of using either the MVT or the MVO core set depends on the situation and the objective. MVO treats breeds as sub-populations of the founder population and estimates the genetic variance of the founder population that survives in the present. While this is a neutral measure of genetic variance, this assumes that all breeds in the set are used to recover the founder by crossbreeding all breeds in the set according to the contributions of each to the core set. This idea may not be very appealing to persons interested in conservation of specific breeds or between breed variance. If this is the case, the MVT core set would be a more logical choice. Applied within a population, where contributions are assigned to individual animals, family groups or sub-lines, the founder variance conserving properties of the MVO core set may be used to maintain genetic variation (chapter 2).

One of the consequences of using core set diversity and kinship measures is the awareness that rare breeds do not necessarily contribute much to the total genetic diversity of a species. By definition a rare breed will be of small (effective) size and have relatively little within breed diversity. If, in addition, such a rare breed is related to other breeds that are larger and more widely used, its contribution to the genetic variance of the species will be (close to) zero. Moreover, based solely on contributions to a core set, large, economically successful breeds will be given priority, while not in (immediate) need of conservation efforts. This emphasises that core set contributions do not provide a definitive answer to the question of what to conserve. There are different criteria and different ways of using genetic diversity measures to prioritise breeds or individuals for conservation. These will be dealt with in the next chapter. But it should always be remembered that (contributions to) genetic diversity is one argument among many others for or against the conservation of a breed.

References

Bennewitz, J. and T.H.E. Meuwissen, 2005. A novel method for the estimation of the relative importance of breeds in order to conserve the total genetic variance. Genetics Selection Evolution 37: 315-337.

Bretting P.K. and M.P. Widerlechner, 1995. Genetic markers and horticultural germplasm management. HortScience 30: 1349-1356.

Caballero A. and M. A. Toro, 2002. Analysis of genetic diversity for the management of conserved subdivided populations. Conservation Genetics 3: 289-299.

Chevalet C., N. Nikolic and M. SanCristobal, 2006. Effects of genetic drift and mutations on some measures of genetic diversity in livestock populations. Proceedings 8th World Congress on Genetics Applied to Livestock Production, CD-ROM Communication No. 33-07.

Eding, H. and T.H.E. Meuwissen, 2001. Marker based estimates of between and within population kinships for the conservation of genetic diversity. Journal of Animal Breeding and Genetics 118: 141–159.

Eding, H., and T.H.E. Meuwissen, 2003. Linear methods to estimate kinships from genetic marker data for the construction of core sets in genetic conservation schemes. Journal of Animal Breeding and Genetics 120: 289–302.

Eding H., R.P.M.A. Crooijmans, M.A.M. Groenen and T.H.E. Meuwissen, 2002. Assessing the contribution of breeds to genetic diversity in conservation schemes. Genetics Selection Evolution 34: 613–633.

Falconer, D.S. and T.F.C. Mackay, 1996. Introduction to Quantitative Genetics. 4th edition, Longman, Harlow, Essex, UK.

FAO, 1998. Secondary Guidelines for Development of National Farm Animal Genetic Resources Management Plans: Management of Small Populations at Risk. FAO, Rome, Italy.

Felius, M., 1995. Cattle breeds of the world. Misset Uitgeverij bv, Doetinchem, The Netherlands.

Felsenstein, J., 1996. PHYLIP, (Phylogeny Inference Package) Version 3.5c, University of Washington, http:\\ evolution.genetics.washington.edu/phylip.html

Frankel, O.H. and A.H.D. Brown, 1984. Plant genetic resources today: a critical appraisal, Crop genetic resources: conservation and evaluation. George Allen & Unwin; London; United Kingdom.

Garcia, D., N. Corral and J. Cañon, 2005. Combining inter- and intrapopulation information with the Weitzman approach to diversity conservation. Journal of Heredity 96: 704-712.

Gilligan, D.M., D.A. Briscoe and R. Frankham, 2005. Comparative losses of quantitative and molecular genetic variation in finite populations of *Drosophila melanogaster*. Genetical Research 85: 47-55.

Katz,M., 1986. Etude des propriétés de certains indices de distance génétique et de leurs estimateurs. Doctoral thesis, Université Paris VII, Paris, France.

Li, C.C., D.E. Weeks and A. Chakravarti, 1993. Similarity of DNA fingerprints due to chance and relatedness. Human Heredity 43: 45-52.

Lynch, M., 1988. Estimation of relatedness by DNA fingerprinting. Molecular Biology and Evolution 5: 584-599.

Lynch, M. and K. Ritland, 1999. Estimation of pairwise relatedness with molecular markers. Genetics 152: 1753–1766.

Malécot, G., 1948. Les Mathématiques de l'Hérédité. Masson, Paris.

Mateus, J.C., H. Eding, M.C.T. Penedo and M.T. Rangel-Figueiredo, 2004. Contributions of Portuguese cattle breeds to genetic diversity using marker-estimated kinships. Animal Genetics 35: 305-313.

Nagylaki, T., 1998. Fixation indices in subdivided populations. Genetics 148: 1325-1332.

Nei, M., 1973. The theory and estimation of genetic distance. Genetic Structure of populations, N.E. Morton University Press of Hawaii.

Nei, M., 1987. Molecular Evolutionary Genetics. Columbia University Press, New York.

Nei, M., F. Tajima and Y. Tateno, 1983. Accuracy of estimated phylogenetic trees from molecular data II, Gene frequency data. Journal of Molecular Evolution 19: 153-170.

Oliehoek, P.A., J. Windig, J.A.M. van Arendonk and P. Bijma, 2006. Estimating relatedness between individuals in general populations with a focus on their use in conservation programs. Genetics 173: 483-496.

Pyasatian, N. and B.P. Kinghorn, 2003. Balancing genetic diversity, genetic merit and population viability in conservation programmes. Journal of Animal Breeding and Genetics 120: 137-149.

Reist-Marti, S.B., H. Simianer, J. Gibson, O. Hanotte and J.E.O. Rege, 2003. Weitzman's appraoch and conservation of breed diversity: an application to African cattle breeds. Conservation Biology 17: 1299-1311.

Reynolds, J., 1983. Estimation of the coancestry coefficient basis for a short-term genetic distance. Genetics 105: 767-779.

Robertson, A. and W.G. Hill, 1984. Deviation from Hardy-Weinberg proportions: Sampling variance and use in estimation of inbreeding coefficients. Genetics 107: 703-718.

Rogers, J., 1972. Measure of genetic similarities and genetic distances. Studies in genetic VII, University of Texas Pub.

Ruane, J., 1999. A critical review of the value of genetic distance studies in conservation of animal genetic resources. Journal of Animal Breeding and Genetics 116: 317-323.

Scherf, B.D., 2000. World Watch List for domestic animals, 3rd edition, FAO, Rome, Italy.

Takezaki, N. and M. Nei, 1996. Genetic distances and reconstruction of phylogenetic trees from microsatellite DNA. Genetics 144: 389-399.

Thaon d'Arnoldi, C., J.-L. Foulley and L. Ollivier, 1998. An overview of the Weitzman approach to diversity, Genetics Selection Evolution 30: 149–161.

Toro, M.A., C. Barragan and C. Ovilo, 2003. Estimation of genetic variability of the founder population in a conservation scheme using microsatellites. Animal Genetics 34: 226-228.

Toro, M.A., C. Barragan, C. Ovilo, J. Rodriganez, C. Rodriguez and L. Silió, 2002. Estimation of coancestry in Iberian pigs using molecular markers, Conservation Genetics 3: 309-320.

Toro, M.A., J. Fernández, and A. Caballero, 2006. Scientific policies in conservation of farm animal genetic resources. Proceedings 8th World Congress on Genetics Applied to Livestock Production, CD-ROM Communication No. 33-05.

Wang, J.L., 2002. An estimator for pairwise relatedness using molecular markers. Genetics 160: 1203–1215.

Weir, B.S., 1990. Genetic data analysis: methods for discrete population genetic data. Sinauer Associates, Inc. Publishers, Sunderland Massachusetts.

Weir, B.S. and C.C. Cockerham, 1984. Estimating F-statistics for the analysis of population structure. Evolution 38: 1358-1370.

Weitzman, M.L., 1992. On diversity. The Quarterly Journal of Economics 107: 363–405.

Wright, S., 1969. Evolution and the genetics of populations, Vol. 2: The Theory of Gene Frequencies. University of Chicago, Chicago, USA.

Chapter 6. Selection of breeds for conservation

Jörn Bennewitz[1], *Herwin Eding*[2], *John Ruane*[3] *and Henner Simianer*[4]
[1]*Institute of Animal Breeding and Husbandry, Christian-Albrechts-University of Kiel, 24098 Kiel, Germany*
[2]*Institute for Animal Breeding, Federal Agricultural Research Centre, Höltystraße 10, 31535 Neustadt, Germany*
[3]*Independent consultant, 49 Via F. Dall'Ongaro, 00152 Rome, Italy*
[4]*Institute of Animal Breeding and Genetics, Georg-August-University Göttingen, 37075 Göttingen, Germany*

Questions that will be answered in this chapter:

- *What are the objectives of a conservation effort?*
- *Which species are relevant?*
- *Is the risk-strategy efficient for the selection of breeds?*
- *Is the maximum-diversity-strategy efficient for the selection of breeds?*
- *Is the maximum-utility-strategy efficient for the selection of breeds?*
- *What are the practical aspects?*

Summary

This chapter presents the different objectives and appropriate strategies for the selection of breeds for conservation. The precise definition of the objective of a conservation effort is crucial for the selection of breeds. Three different selection strategies are presented and discussed. The frequently applied maximum-risk-strategy, which uses basically the degree of endangerment of a breed as the sole selection criteria, is sub-optimal because it does not follow a well defined objective. The maximum-diversity-strategy is efficient, if safeguarding genetic diversity to maintain flexibility in the future is the objective. If a more comprehensive objective is defined that includes also utilisation (like presenting the cultural values of breeds), then the maximum-utility-strategy can be applied, which is an extension of the maximum-diversity-strategy. The chapter ends with some practical aspects that should be considered when selecting breeds for conservation.

1. Introduction

On a world-wide level, there are around 7,600 breeds from 35 domestic mammals and bird species. Around 30% of them are classified either as endangered, at risk or even extinct (Scherf *et al.*, 2006). Additionally, from around 35% of the breeds the classification status is unknown, but it can reasonably be assumed that a fair proportion of these breeds is also endangered. The breed extinction process is different for the species (chapter 1) but no farm animal species is endangered as such. On the other hand, the financial funds available (either private or public) for conservation of farm animal biodiversity are limited, preventing conservation of all endangered breeds. Indeed, it is not possible and also may not be desirable to conserve all endangered breeds. Many breeds are members of breed groups and can be replaced by other breeds without losing biodiversity. Other highly endangered breeds with a few animals left may be genetically impoverished so that efforts to maintain them may not represent a cost-efficient contribution to biodiversity (Ruane, 2000).

The choice of breeds that should be included in a well designed and efficient conservation plan is of fundamental importance. In the following several strategies for the selection of breeds for conservation are reviewed and discussed with respect to the objectives of the conservation plan.

2. What are the objectives of a conservation effort?

The arguments for the conservation of farm animal biodiversity are outlined in detail in chapter 2 and can be classified either into insurance arguments or into sustainable utilisation of rural area arguments. The insurance arguments imply conservation of sufficient genetic diversity in order to be able to cope with putative future changes in the production or market environment. This includes conserving breeds which show some special traits that are of interest (examples in Box 6.1). Here it seems that at least one breed should be conserved that shows this special trait. Additionally, breeds which contribute to the genetic diversity of the species should be conserved. The insurance argument focuses on the maintenance of diversity within species.

The specific adaptation of a breed to its, sometimes very harsh, production environment falls into the sustainable utilisation argument. This specific adaptation is a consequence of natural and artificial selection and mating of animals according to a breeding goal (either well defined or somewhat intuitive) in order to form a breed that copes with the particular challenge of the environment (examples in Box 6.1). Other breeds show a long history, parallel with the cultural development of human populations. Hence, in the same way as other cultural assets like old buildings or artwork, some breeds can be considered as cultural and historical merits and should therefore be conserved. The sustainable utilisation argument favours the *in situ* conservation of specific breeds in their present state.

Box 6.1. Examples of well adapted breeds and of breeds with unique traits.

- The Kuri cattle's adaptation to the aquatic environment around the island and shores of the Lake Chad Basin is one of the clearest examples of adaptation to a specific environment (Tawah *et al.*, 1997).
- The feral Soay sheep is adapted to the very tough conditions of the St. Kilda archipelago, off the coast of Scotland, and is thought to have existed there since the Neolithic times (Hall and Bradley, 1995).
- The Meishan pig breed from China is extremely fertile and has thus been imported to Europe and North America for use in commercial stocks, as well as for study in a wide range of research projects to understand the genetic basis of fertility, and indeed, of other traits (e.g. Janss *et al.*, 1997).
- Study of the unique musculature of the Belgian Blue cattle breed has led to an increased understanding of the genetic mechanism behind muscular development in mammals (Grobet *et al.*, 1997).
- The North Ronaldsay sheep breed is unique in that it feeds only on seaweed for large parts of the year and, in addition, has very efficient copper absorption and high salt tolerance (Ponzoni, 1997).
- The rare Gulf Coast native sheep has high natural resistance to internal parasites, a characteristic which led to flocks of the breed being established at the Universities of Florida and Louisiana for research purpose (Henson, 1990).
- Many other breeds have documented resistance to specific diseases, such as the N'dama cattle which survive in areas infested by the tsetse fly, due to their high resistance to trypanosomiasis (the cattle equivalent to sleeping sickness, transmitted by the fly), and have been at the centre of a large research programme (FAO, 1992).

Examples of well adapted breeds and of breeds with unique traits reported to FAO in the individual country reports within the State of the World's process, SOW (chapter 1):

- Within the Icelandic Sheep population a line with leadership characteristics exists. It is an unusual ability to find their way and they heave greater intelligence than other sheep lines. Crossbreeding with the Leader-line shows a clear heritability for this trait (Icelandic Ministry of Agriculture, 2003).
- In Zambia the Angoni, Barotse, Baila and Tonga cattle breeds are adapted to the local tropical conditions and are heat, parasite and disease tolerant (Farm Animal Genetic Regional Focal Point of Zambia, 2003).
- In the Republic of Kazakhstan the Kazakh fine-flees sheep breed is famous for the wool quality and quantity (Ministry of Agriculture of the Republic of Kazakhstan, 2003).
- In the Republic of Korea the Yeonsan Ogol chicken breed is totally black including the the muscle, the bone and the intestinal organs. It has been raised for several hundreds of years and is used for medicinal purpose (Republic of Korea, 2004).

The precise definition of the objectives for conservation is crucial for the selection of breeds to be included in the conservation plan. If the objective of the conservation effort is only to maintain as much neutral genetic diversity as possible, then only those breeds which contribute most to it should be selected and *ex situ* conservation might be sufficient. If, however, the objective includes also additional objectives for sustainable use in rural areas, then also those breeds that fulfil the additional criteria are relevant for the conservation plan including *in situ* conservation (chapter 2).

3. Which species are relevant?

The different domesticated species carry out a multitude of purposes, providing humans with food (meat, milk, and eggs), fibre, leather, transport and, as by-products, fertiliser and fuel. Some species are more important than others, e.g. almost all the world's milk comes from cattle and buffaloes and the world's meat from pigs, chicken and cattle. There are also domesticated species that are important in the developed world as companion animals like cats and dogs.

For the selection of breeds across species, two equal-strategies based on some form of uniform allocation may be considered. The first assigns a specific amount of financial resources to each species. This would result in a larger number of conservation programmes for smaller less-costly species like poultry or rabbits. The second considers an equal number of breeds per species. Here the majority resources would be spent for larger more-costly species like cattle or horses. These equal approaches are not convincing, because they do not take the relative importance of the species into account.

The prioritisation of species should be done in accordance with the respective objective of the conservation plan. If the plan includes also the conservation of companion animals, perhaps because of cultural arguments, then their conservation should also receive support and breeds from these species have to be selected. If the focus is only on species with agricultural relevance, then companion animals should be excluded. In this case those species with many endangered breeds that are important for agricultural production should receive most support.

4. Is the risk-strategy efficient for the selection of breeds?

At the moment, the often applied rules by the stakeholders for selection of breeds for conservation mostly rely on a single or a combination of a few simple criteria, which are related to the risk status of a breed. Risk status is deduced from the number of breeding males and females, the inbreeding rate (estimated from the effective population size, chapter 3) or population dynamics like increasing or decreasing population size.

Such schemes may have different steps, putting a breed on a 'watch' status when the risk-describing parameter reaches a defined threshold value, and starting specific and previously defined activities if it falls below further thresholds. In Tables 6.1 and 6.2 the risk categories as defined by FAO (Scherf, 2000) and the European Association of Animal Production (EAAP, 1998) are shown, respectively. A framework to uniform risk categories across institutions has been proposed (Gandini *et al.*, 2005).

Although simple and pragmatic, almost all risk-strategies lack criteria that characterise a rational and cost effective decision making process from a systematic viewpoint. The main criticism is that the objective of such conservation strategy is not well defined. Here the implicit objective (implicit, because it is not defined as such) is to conserve all existing breeds. The specific value of the breed (addressing the sustainable utilisation argument for conservation) or its contribution to the genetic diversity (the insurance argument) is not directly accounted for. Hence, the risk-strategy in itself is not efficient in selection of breeds for conservation.

Table 6.1. Risk categories used by FAO (Scherf, 2000).

Risk category	Number of					Additional criteria
	females		males		total breeding animals	
Extinct	0	or	0			Impossible to re-establish the breed
Critical	< 100	or	< 5	or	< 120 and decreasing and < 80% pure breeding	
Critical - maintained						Critical + conservation or commercial breeding program in place
Endangered	< 1000	or	< 20	or	between 80 and 100 and increasing and > 80% pure breeding	
				or	between 1,000 and 1,200 and decreasing and < 80% pure breeding	
Endangered - maintained						Endangered + conservation or commercial breeding program in place
Not at risk	> 1000	and	> 20	or	> 1,200 and increasing	Other categories don't apply

Table 6.2. Risk categories based on inbreeding rate (ΔF) used by the EAAP (1998).

Risk category	ΔF over 50 years
Critically endangered	> 40%
Endangered	26 – 40%
Modestly endangered	16 – 25%
Possibly endangered	5 – 15%
Not endangered	< 5%

Note: The following additional elements are used to adjust the risk category by one category: proportion of registration in the herd book, change in the number of breeding animals, percentage of pure breeding and immigration and number of herds.

Ruane (2000) suggested a framework for prioritising breeds for conservation on the national level and applied this to 45 Norwegian breeds covering 17 species. He scored breeds for the following criteria: degree of endangerment (mainly determined by the current population size plus some additional factors), traits of current economic value, special landscape value, traits of current scientific value, cultural and historical value and genetic uniqueness. He suggested to use the scoring list for the selection of breeds for conservation. He did not propose a general algorithm for combining scores for the different criteria, arguing that this would depend on country-specific conditions and priorities. He suggested nevertheless that the degree of endangerment should be the most important criterion.

5. Is the maximum-diversity-strategy efficient for the selection of breeds?

Using the maximum-diversity-strategy, a breed is selected for conservation that contributes significantly to the genetic diversity. The prerequisite for this approach is that an appropriate diversity measure is used that reflects the objective of the conservation program. In chapter five different diversity measures were described, that differ from their conceptual point of view as well as in the weighing of the between- and within-breed diversity. In the following it will be shown how diversity measures can be used for selection of breeds for conservation. For the relative importance of the between- and within-breed diversity, and hence the choice of the appropriate diversity measure for the selection of breed, see Box 6.2.

Following the maximum-diversity-strategy, breeds can be ranked according to their contribution either to the actual or to the expected future diversity. The drawback of ranking breeds according to their contribution to the actual diversity is that the loss

Box 6.2: Relative importance of between and within breed genetic diversity.

In chapter 3 it was shown how diversity can be partitioned into between-breed diversity (D_B) and within-breed diversity (D_w). The question is, which component is more relevant for conservation and hence for the selection of breeds for a conservation plan. Fabuel *et al.* (2004) investigated the impact of a different weighing factor λ in the combination of the two components, i.e. $\lambda D_w + D_B$, and obtained different breed conservation priorities, depending on λ. Following this, the determination of the relative importance of these two components is crucial and it determines the choice of the diversity measure (chapter 5) that has to be used in the maximum-diversity strategy for the selection of breeds. For instance, if only the between-breed diversity shall be considered ($\lambda = 0$), the Weitzman diversity measure might be appropriate. However, if both between- and within-breed diversity shall be taken into account, the core set diversities maybe used. The Maximum-Variance-Total core set diversity measure maximises the variance of a hypothetical quantitative trait and gives more weight to the between-breed diversity ($\lambda = 0.5$, as pointed out by Toro *et al.* 2006) than the Maximum-Variance-Offspring ($\lambda = 1$) does.
A stronger weighing of the between-breed diversity might be appropriate when conserved breeds are considered to be used in a crossbreeding plan, because both heterosis and complementarity are a function of this type of diversity (Fabuel *et al.* 2004). Additionally, as argued by Piyasatian and Kinghorn (2003) and Bennewitz *et al.* (2006), the between-breed diversity is more accessible, because of the specific gene and allele combination that can be found in different breeds. This allows a faster adaptation of commercial breeds to changes in the production environment or in market needs by introducing genetics from a conserved breed. However, a too strong weighing of between-breed diversity results in ignoring most of the total diversity. The within-breed diversity might be important for the creation of a new synthetic breed that copes with a challenging environment. On the other hand, an over-emphasis on within-breed diversity will favour large breeds, which are not endangered. Additionally, the better accessibility of the between-breed diversity is not accounted for.
In summary, it seems that a diversity measure should be used that considers both components of diversity with a weight that is not arbitrarily chosen by the researcher but is given by the definition of the appropriate diversity measure and chosen according to a well defined conservation objective.

of between breed and within breed diversity over time due to extinction of breeds and genetic drift is not accounted for. Hence, it seems to be better to rank the breeds according to their contribution to the expected future diversity, as it will be shown next.

Assume that there are a number of N breeds included in the analysis and for each breed the probability is known that it will go extinct within a defined future time horizon t (e.g. 50 years), the extinction probability z. At the end of the defined time horizon there are 2^N different possibilities of present and extinct breed combinations possible, i.e. there are 2^N different breeds sets K, each with a certain probability P_K, which depends solely on the extinction probabilities of the breeds. Each breed set K shows a diversity D_K. Following this, the expected conserved diversity at the end of the time horizon t is

$$E(D_t) = \sum_K P(K)D_K ,$$

which is a projection of the actual diversity into the future. It depends on the extinction probability z (and on the expected drift, if the loss of within breed diversity is accounted for; see Simianer, 2005b; Bennewitz and Meuwissen, 2006). The effect of a breed could be assessed by how much the expected diversity would be changed with respect to a small change in the breed extinction probability. This is essentially the concept of marginal diversities. A marginal diversity of a breed (md_i) is defined as the change of conserved diversity at the end of the considered time horizon, when the extinction probability of the breed, z_i, would be changed by one unit by a conservation effort (Weitzman, 1993). It can be estimated as the first partial derivative of $E(D_t)$ with respect to z_i, i.e.

$$md_i = -\partial E(D_t) / \partial z_i .$$

The minus sign makes it positive, i.e. it reflects the change of the conserved diversity, when the extinction probability would be lowered by one unit. The estimation of marginal diversities can be done with any diversity measure that fulfils the 'monotonicity' and 'non-negativity' properties of the Weitzman criteria for a proper diversity measure (chapter 5). For computational details see Simianer *et al.* (2003) and Bennewitz *et al.* (2006). It is important to note that the marginal diversity of a particular breed is independent of its extinction probability.

The marginal diversities can be multiplied by the extinction probabilities in order to obtain the conservation potentials of the breeds, i.e. $CP_i = md_i * z_i$ (Weitzman, 1993). The conservation potential gives an idea how much diversity can be conserved additionally if a particular breed would be made completely safe. It was shown that the maximum-diversity-strategy using the conservation potentials for prioritising breeds is very efficient in selection of breeds for conservation, when diversity is the objective of the conservation plan (Reist-Marti *et al.*, 2003, Simianer *et al.*, 2003). If, however, also other arguments are included, the maximum-diversity-strategy has to be extended to the maximum-utility-strategy as it will be shown in the next section.

The difficulties in this approach are that extinction probability estimates are required for the breeds, and these are not easy to obtain. In Box 6.3 the difficulties in their estimation are described. However, it was observed that the marginal diversities are not very sensitive to changes in extinction probabilities and that these have to be known

Box 6.3. The extinction probabilities of breeds.

The extinction probability of a breed is defined as the probability that a breed will go extinct at some point within a defined future time horizon (e.g. 25 or 50 years). The problem in modelling, and consequently in estimating these probabilities, is that extinction of a breed is a rare event and therefore any model validation and formal model comparison is almost impossible.
A semi-quantitative method was applied to a set of 49 African breeds by Reist-Marti *et al.* (2003). These authors scored the breeds for four variables related to the population (population size and its change over time, distribution of the breed and risk of discriminate crossing), four related to the environment (organisation among farmers, existence of a conservation scheme, political situation and reliability of the information) and two related to the value of the breeds (presence of special traits and cultural value). The extinction probabilities of the breeds were calculated as the sum of the 10 variables and were re-scaled to a value between 0.1 and 0.9 in order to prevent extreme probabilities. Probabilities of zero and one were not allowed, because the future cannot be foreseen. This approach is appealing, because of its comprehensiveness.
Simianer (2005b) argued that the extinction probability of a breed is directly related to the rate of inbreeding. Following this, he obtained extinction probabilities as $1/2N_e$ and multiplied them by a constant to obtain reasonable values. Therefore these probabilities can be interpreted as relative rather than absolute probabilities. The same holds true for the estimates obtained from the Reist-Marti method. The problem of these two methods is that they do not produce any standard errors or confidence intervals of the extinction probabilities.
A quantitative method adapted from conservation biologists was used by Bennewitz and Meuwissen (2005). This method is based on a time series approach and involves a random process to predict likely future population size based on recent census data. On the one hand, the method produces absolute rather than relative extinction probabilities and also confidence intervals for the probabilities. However, the extinction probabilities were either close to zero or close to one and the confidence intervals covered almost the whole parameter space. The reason may be, that this method is tailored to wildlife populations, which show much greater amplitudes in population size over time.

only to proportionality (Bennewitz *et al.*, 2006). When, in addition, the loss of within breed diversity will be considered, estimates of the effective population size are required for the quantification of the expected drift. It should be noted that the breeds selected for conservation based on their conservation potentials are not necessarily the most endangered breeds, making the results of the maximum-diversity-strategy different from the risk-strategy. For example, the correlation between the extinction probabilities and conservation potentials was only around 0.4 in the study of Bennewitz *et al.* (2006), involving 44 North Eurasian cattle breeds. In Box 6.4 an example of the results of the maximum-diversity-strategy is presented.

Box 6.4. Example of the application of the maximum-diversity-strategy.

Bennewitz and Meuwissen (2006) used a small data set consisting of nine Dutch cattle breeds genotyped for a number of microsatellite markers to demonstrate the maximum-diversity-strategy. They used the Maximum-Variance-Total core set diversity measure (chapter 5). The marginal diversities (*md*) considered the expected loss of between breed diversity due to extinction of breeds and the expected loss of within breed diversity due to drift. The prioritising of breeds for conservation could be done accordingly to the conservation potentials (*CP*).

Breed	Effective population size[1]	Extinction probability	md_i[2]	CP_i[3]
Belgian Blue	370	0.027	11.14	0.301
Dutch Red Pied	68	0.147	34.17	5.023
Dutch Black Belted	154	0.065	32.22	2.094
Limousine	400	0.025	154.74	3.869
Holstein Friesian	>1000	0.001	23.61	0.024
Galloway	23	0.435	190.48	82.859
Dutch Friesian	294	0.034	5.82	0.198
Improved Red Pied	111	0.090	41.77	3.759
Blonde d'Aquitaine	217	0.046	8.31	0.382

[1]The effective population sizes were taken from the database of the EAAP.
[2]*md* describes how much the expected future diversity would change with respect to a small reduction in a breed's extinction probability.
[3]*CP* describes how much the expected future diversity would change if a breed was made completely safe, i.e. $CP = md \times$ extinction probability.

Although conservation potentials are very useful for prioritising, they do not tell us anything about the optimal allocation of the budget with respect to maximising the conserved diversity. For the optimal allocation of the budget, the analysis has to involve a cost function for the reduction of extinction probabilities of the breeds (Weitzman, 1993; Simianer, 2002; Simianer *et al.*, 2003). More precisely, this method assumes that marginal costs and marginal returns (in diversity) of conservation activities can be specified for each breed. The total budget available for conservation is then allocated over the selected breeds using an iterative algorithm in order to maximise the conserved diversity. See Box 6.5 for further details of the optimum allocation approach.

An alternative to the use of conservation potentials is the so-called 'safe set+1' approach as used by Thaon d'Arnoldi *et al.* (1998) and Eding *et al.* (2002). Following this, a safe

Box 6.5. The optimum allocation of financial funds over selected breeds for conservation.

The optimal allocation approach proposed by Simianer (2002) and Simianer *et al.* (2003) makes the following assumptions: The financial funds available for conservation is most efficiently used if the expected diversity at the end of the considered time horizon t (e.g. $t = 25$ years) is maximised. By investing a certain share of the available resources in breed i, the extinction probability z_i of this breed will be changed to $z_i^* < z_i$, resulting in an increase in the expected diversity at t, i.e. $E^*(D_t) > E(D_t)$. The conservation effect $\Delta z_i = z_i^* - z_i < 0$ is a function of both the extinction probability of the breed and the amount of funds invested in the conservation of this breed. This cost function $\Delta z_i = f(z_i, b_i)$ has to be specified. Now, let $B = \{b_i\}$ be a vector describing a fixed pattern of allocation of funds to a set of breeds. For each breed with $b_i > 0$, the resulting change in extinction probability can be computed using the specified cost function, $\Delta z_i = f(z_i, b_i)$ and a new $E^*(D_t) > E(D_t)$ can be calculated using the reduced extinction probabilities $z_i^* < z_i$. This increase of expected diversity is the expected effect of the allocated funds.
Under the assumption made above, the optimum allocation of funds can be found using the following algorithm. Divide the total fund into n_b equal and small shares of money β. Then follow the iterative procedure:

1. Set $b_i = 0$ for al breeds and start with the first share β.
2. Compute the expected reduction of extinction probability Δz_i for each of the breeds under the assumption that β is spent on only this breed.
3. Compute the increase in expected diversity $E(\Delta D_t \mid z_i, \beta) = \Delta z_i md_i$ for each breed, where md_i is the marginal diversity of the breed.
4. Allocate this share on breed j, for which the increase of expected diversity is highest; update the extinction probability of this breed from the actual value z_j by Δz_j to z_j^* and add β to b_j.
5. Recalculate marginal diversities for all breeds.
6. Allocate the next share, beginning with step 2, until all shares are allocated.

After going through the described iterative algorithm, the vector B contains the optimal allocation of the available funding to the set of breeds in the sense that no other pattern of allocation would lead to a higher quantity of conserved diversity.
One of the difficulties of this approach is the specification of the cost function. Based on arguments from population genetics, Simianer *et al.* (2003) suggested three types of cost functions, which may reflect the range of possible functions in typical conservation situations. The authors applied this method to 23 African zebu and zenga cattle breeds, using extinction probabilities of Reist-Marti *et al.* (2003) (Box 6.3) and the Weitzman diversity measure (chapter 5). They found that conservation funds should be spent on only three to nine of the breeds with different proportions, depending on the used cost function. Highest amount of funds should be given to those breeds that show a large conservation potential.
The optimum allocation approach can also be applied using the marginal utilities of the breeds (paragraph 6) instead of the marginal diversities. This would allow considering

▷▷▷

also other features included in the conservation objective, e.g. special traits, fixed and variable cost of conservation schemes etc. See Reist-Marti *et al.* (2006) for an application.

set is formed by breeds that can be considered safe from extinction in the near future. This may encompass breeds that are currently widely used or breeds that already are (or will definitely be) subject to conservation (due to special traits, etc.). The diversity is estimated that is conserved by these breeds. Then the breeds not in the safe set are added one by one with re-placement to the safe set and the increase in conserved diversity of the safe set+1 is estimated. Those breeds that cause the largest increase in conserved diversity obtain higher priority in the conservation plan. The advantage of this simple approach is that no extinction probabilities need to be specified; only the breeds for the safe set have to be chosen. This, however, can also be seen as the biggest disadvantage, since the choice of breeds for conservation is totally independent of the breed's degree of endangerment.

6. Is the maximum-utility-strategy efficient for the selection of breeds?

As mentioned in the previous section, the maximum-diversity-strategy is efficient if diversity is the only objective of a conservation plan. If, however, also other features are included (sustainable use in rural areas; chapter 2) in the objective, the maximum-diversity-strategy can be straightforwardly extended to the maximum-utility-strategy, as it will be shown next. For further details, including applications, the interested reader is referred to Simianer (2002), Simianer *et al.* (2003) and Reist-Marti *et al.* (2006). These authors also described the use of the optimum allocation scheme (Box 6.5) in combination with the utility. The idea of the use of the utility was first raised by Weitzman (1998).

Let us assume that the objective of a conservation plan includes both the sustainable utilisation and the insurance arguments, the latter including neutral diversity and special traits (paragraph 2). In this case the utility conserved by a set of non-extinct breeds (denoted by K, as in the previous section) at a defined future time horizon can be written as (Simianer *et al.*, 2003)

$$U_K = w_D D_K + \sum_{j \in K} w_{F_j} + \sum_i k_i w_{B_i} ,$$

where

- U_K is the utility of the breed set K,
- w_D is the relative value of a unit neutral diversity,
- D_K is the neutral diversity of the breed set K;
- w_{F_j} is the relative value of feature j (e.g. a special trait) and $j \in K$ denotes for feature j being present in at least one of the non-extinct breeds, i.e. present in the set K;

- w_{B_i} is the relative value of breed i (the sustainable utilisation value of the breed);
- k_i is an indicator variable that is equal to 1 if breed i is present in the set K (i.e. not extinct) or zero otherwise (i.e. extinct at the end of the time horizon), depending on its extinction probability.

The first two terms of the equation shown above address the insurance arguments and the last term the sustainable utilisation argument. As already stated above, there are 2^N different breed sets K possible, each with a certain probability P_K. The expected conserved utility at the end of the time horizon t is

$$E(U_t) = \sum_{\forall K} P(K)U_K \; .$$

Now a marginal utility of a breed can be estimated as

$$mu_i = -\delta E(U_t)/\delta z_i,$$

which is similar to the estimation of the marginal diversities shown above. Here, a marginal utility is defined as the change in conserved utility at the end of the defined time horizon when the extinction probability would be lowered by one unit by a conservation effort. Similarly, a conservation potential of a breed with respect to the utility can be estimated as the product of the marginal utilities and the extinction probabilities and these can be used to select breeds for conservation.

The second term in the equation shown above, the marginal utility with respect to the special features like e.g. special traits, deserves attention. From a conservation point of view, it is desirable to maintain such features by conserving at least one breed which has the respective feature. Consequently, the marginal utility of a breed in this context is heavily dependant on the composition of the set of breeds. For example, if a special trait is present in a number of breeds, of which one is almost perfectly safe, the marginal utility of the other breeds is low or even zero. If, however, only one breed is left with the desired trait, its marginal utility will be very high, because when it goes extinct, the special trait will be lost for the whole species.

From a conceptual and systematic point of view, the maximum-utility-strategy seems to be the most favourable method for the selection of breeds. It reduces to the maximum-diversity-strategy if diversity alone is in the objective of the conservation plan. The problem with this approach is that next to the need for estimates for extinction probabilities (as in the maximum-diversity-strategy) additional estimates for the relative economic values of neutral diversity, of the special features (e.g. special traits) and for the breed specific values (e.g. the historical value of a breed) are needed (w_D, w_{F_j}

and w_{F_i}, respectively). At the moment there is no obvious way to obtain these relative weights. The first attempts have been made by Gandini and Villa (2003) to determine the cultural values of breeds (chapter 2). Definitely more research is needed to obtain the economic weights for getting the full benefits out of the maximum-utility-approach.

7. What are the practical aspects?

Selection of breeds for conservation is always faced with a substantial amount of uncertainty. We usually have good knowledge about the inventory of breeds and their relationship to each other, the latter one often assessed with the aid of genetic markers (see chapter 5). However, as described in the previous section, if it comes to extinction probabilities, or even to economic weights of conservation arguments, the uncertainty reaches a considerable level. Having this in mind, it seems advisable that the selection of breeds for conservation should be made according to the defined objective of the conservation effort. The maximum-utility-strategy is appealing because it is possible to include all arguments of the objective. However the level of uncertainty reaches its maximum when applying this strategy. This does not mean that simpler strategies like the risk-strategy are to be preferred. The latter one just ignores the level of uncertainty in the way that it uses a not well defined objective of conservation. Considerable research is needed to reduce the level of uncertainty, mainly in the estimation of the relative economic weights of the conservation arguments. In the mean time somewhat intuitive estimates or best guesses could be used in the maximum-utility-strategy. For example, Reist-Marti *et al.* (2006) included in their utility-function neutral diversity and eight different special traits, always carried by several, but not all, breeds. They derived relative weights for special traits in the sense that they could quantify how much conserved expected diversity needs to be sacrificed to conserve a special trait by a given magnitude. Although this does not solve the problem of putting economic weights on the same scale to diversity and special traits, respectively, this approach provides a relative valuation of these two different quantities. Other simplified approaches are thinkable. A sensitivity analysis could serve an idea how the results would change, when different economic weights would be used.

One argument frequently raised by the stakeholders is that it is much more difficult to implement other selection rules than the risk-strategy, because the risk-strategy is easy to communicate to people from breeding organisations or farmers. However, since the selection of breeds is of fundamental importance for an efficient conservation effort, the best available strategy should be used for it.

A very important aspect that has finally to be mentioned is that decisions are made on the appropriate scale (Simianer, 2005a). Decision making at the national level is important, where e.g. the responsibilities of countries to their own genetic resources are outlined in the Convention on Biological Diverstiy, a legally-binding agreement with 188 parties. In addition, because livestock breeds and species are spread across political borders, decisions should also be made at the regional or even global level, which is one of FAO's objectives, and conservation activities should be documented and co-ordinated on the international level.

References

Bennewitz J. and T.H.E. Meuwissen, 2005. Estimation of extinction probabilities of five German cattle breeds by population viability analysis, Journal of Dairy Science 88: 2949-2961.

Bennewitz, J., J. Kantanen, I. Tapio, M.H. Li, E. Kalm, J. Vilkki, I. Ammosov, Z. Ivanova, T. Kiselyova, R. Popov and T.H. E. Meuwissen, 2006. Estimation of breed contributions to present and future genetic diversity of 44 North Eurasian cattle breeds using core set diversity measures. Genetics Selection Evolution 38: 201-220.

Bennewitz J. and T.H.E. Meuwissen, 2006. Breed conservation priorities derived from contributions to the total future genetic variance. Proceedings 8th World Congress on Genetics Applied to Livestock Production, CD-Rom Communication No. 33-06.

Eding, J.H., R.P.M.A Crooijmans, M.A.M. Groenen and T.H.E. Meuwissen, 2002. Assessing the contribution of breeds to genetic diversity in conservation schemes. Genetics Selection Evolution 34: 613-633.

EAAP, 1998. Assessment of the degree of endangerment of livestock breeds. Working group on Animal Genetic Resources, 49th Annual Meeting European Association of Animal Production, Warsaw 1998.

Fabuel, E., C. Barragán, L. Silió, M.C. Rodriguez and M.A. Toro, 2004. Analysis of genetic diversity and conservation priorities in Iberian pigs based on microsatellite markers. Heredity 93: 104-113.

FAO, 1992. Proceedings of an expert consultation on 'Programme for the control of African animal trypanosomiasis and related development', Animal Production and Health Paper 100, FAO, Rome, Italy.

Farm Animal Genetic Regional Focal Point of Zambia, 2003. SOW report Zambia 2003. Farm Animal Genetic Regional Focal Point of Zambia.

Gandini, G.C. and Villa, E., 2003. Analysis of the cultural value of local livestock breeds: a methodology. Journal of Animal Breeding and Genetics 120: 1-11.

Gandini G.C., L. Ollivier, B. Danell, O. Distl, A. Georgudis, E. Groeneveld, E. Martiniuk, J. van Arendonk and J. Woolliams, 2005. Criteria to assess the degree of endangerment of livestock breeds in Europe. Livestock Production Science 91: 173-182.

Grobet, L., L.J.R. Martin, D. Poncelet, D. Pirottin, B. Brouwers, J. Riquet, A. Schoeberlein, S. Dunner, F. Menissier, J. Massabanda, R. Fries, E. Hanset and M. Georges, 1997. A deletion in the bovine myostatin gene causes the double-muscled phenotype in cattle. Nature Genetics 17: 71-74.

Hall, S.J.G. and D.G. Bradley, 1995. Conserving livestock breed diversity. Trends in Ecology and Evolution 10: 267-270.

Henson, E.L., 1990. The organisation of live animal preservation programmes. FAO Animal Production and Health Paper 80: 103-117.

Icelandic Ministry of Agriculture, 2003. SOW report Iceland 2003. Icelandic Ministry of Agriculture.

Janss, L.L.G., J.A.M. van Arendonk and E.W. Brascamp, 1997. Segregation analyses for presence of major genes affecting growth, backfat and litter size in Dutch Meishan crossbreds. Journal of Animal Science 75: 2864-2876.

Ministry of Agriculture of the Republic of Kazakhstan, 2003. SOW report 2003 Kazakhstan. Ministry of Agriculture of the Republic of Kazakhstan.

Ponzoni, R.W., 1997. Genetic resources and conservation, Chapter 16. In: The Genetics of Sheeps, L. Piper and A. Ruvinski (eds.), CAB International, pp. 437-469.

Piyasatian N., and B.P. Kinghorn, 2003. Balancing genetic diversity, genetic merit and population viability in conservation programmes. Journal of Animal Breeding and Genetics 120: 137-149.

Reist-Marti, S.B., H. Simianer, J. Gibson, O. Hanotte and J.E.O. Rege, 2003. Weitzman's appraoch and conservation of breed diversity: an application to African cattle breeds. Conservation Biology 17: 1299-1311.

Reist-Marti, S.B., A. Abdulai and H. Simianer, 2006. Optimum allocation of conservation funds and choice of conservation programs for a set of African cattle breeds. Genetics Selection Evolution 38: 99-126.

Republic of Korea, 2004. SOW report 2004. Republic of Korea.

Ruane, J., 2000. A framework for prioritizing domestic animal breeds for conservation purposes at the national level: a Norwegian case study. Conservation Biology 14: 1385-1393.

Scherf, B.D. (ed.), 2000. World watch list for domestic animal diversity. 3rd edition, Rome, FAO.

Scherf, B., B. Rischkowsky, D. Pilling and I. Hoffmann. 2006. The state of the world's animal genetic resources. Proceedings 8th World Congress on Genetics Applied to Livestock Production, CD-Rom Communication No. 33-13.

Simianer, H., 2002. Noah's dilemma: which breeds to take aboard the ark? Proceedings 7th World Congress on Genetics Applied to Livestock Production, CD-Rom Communication No. 26-02.

Simianer, H., S.B. Reist-Marti, J. Gibson, O. Hanotte, and J.E.O. Rege, 2003. An approach to the optimal allocation of conservation funds to minimise loss of genetic diversity between livestock breeds. Ecological Economics 45: 377-392.

Simianer, H., 2005a. Decision making in livestock conservation. Ecological Economics 54: 559-572.

Simianer, H., 2005b. Using expected allele number as objective function to design between and within breed conservation of farm animal biodiversity. Journal of Animal Breeding and Genetics 122: 177-187.

Toro, M.A., J. Fernández, and A. Caballero, 2006. Scientific policies in conservation of farm animal genetic resources. Proceedings 8th World Congress on Genetics Applied to Livestock Production, CD-ROM Communication No. 33-05

Tawah, C.L., J.E.O. Rege, and G.S. Aboagye, 1997. A close look at a rare African breed – the Kuri cattle of Lake Chad Basin: origin, distribution, production and adaptive characteristics. South African Journal of Animal Science 27: 31-40.

Thaon d'Arnoldi, C., J.-L. Foulley, and L. Ollivier, 1998. An overview of the Weitzman approach to diversity. Genetics Selection Evolution 30: 149-161.

Weitzman, M.L., 1993. What to preserve? An application of diversity theory to crane conservation. The Quarterly Journal of Economics CVII: 157-183.

Weitzman, M.L., 1998. The Noah's ark problem. Econometrica 66: 1279-1298.

Chapter 7. Genetic contributions and inbreeding

John Woolliams[1,2]
[1]*Roslin Institute, Roslin Midlothian EH25 9PS, United Kingdom*
[2]*Department of Animal and Aquacultural Sciences, Norwegian University of Life Science, P.O.Box 1432, Ås, Norway*

Questions that will be answered in this chapter:

- *How do we recognise inbreeding and how do we measure it?*
- *What determines how fast inbreeding accumulates?*
- *What principles can be adopted to manage inbreeding?*
- *How do these principles differ between breeding schemes primarily concerned with conservation and those primarily concerned with breed improvement?*
- *How might rates of inbreeding be predicted?*
- *What general recommendations can be made for size of breeding schemes to achieve a commonly accepted minimum Ne of 50?*

Summary

This chapter examines how and why inbreeding accumulates in a population, and introduces the concept of genetic contribution as a simple means of describing the rate of inbreeding (ΔF). The relationship between ΔF and genetic contributions is used to derive the basic principles of managing a population to minimise ΔF. A further relationship between the genetic contributions and the rate of gain in a selected trait is developed to illustrate how the relationships between long term genetic contributions and rates of gain and inbreeding can lead to predictions on the optimum way to select for genetic improvement whilst managing the rate of loss of genetic variation. Approaches to predicting ΔF are examined, and guidelines are given on ways to manage ΔF in practice depending on the sophistication of the breeding scheme. These principles will be developed further for practical implementation in chapter 8.

1. Inbreeding and inbreeding rate

Inbreeding was defined in chapter 3 and related aspects have been developed in previous chapters to identify resources for establishing conservation schemes and for uncovering aspects of the history of the population. However the management of a genetic resource requires procedures for sustaining current diversity, including awareness of practices that may unnecessarily increase inbreeding. Therefore this chapter will discuss the concept of genetic contributions and how this simple concept can be used to inform thinking.

Inbreeding coefficients are used extensively in the descriptions of populations, and those managing populations frequently ask 'what is a safe level of inbreeding?', particularly when it is noted that inbreeding coefficients are all >0 in their population, with many no longer close to 0. However this is not an appropriate question for the following reasons:

1. Inbreeding is inevitable. Inbreeding is >0 when an individual has two ancestors in common and to avoid inbreeding, it is necessary to have 2 parents, 4 distinct grandparents, 8 distinct great-grandparents and so on. The number of distinct ancestors required very quickly exceeds the population size in the past.
2. In a closed population inbreeding will increase and will eventually exceed any pre-determined value of F. Lost alleles cannot be replaced without migration from another gene pool.
3. Each generation new mutations enter the population and these add to the population genetic variance, compensating in part for that variation lost by inbreeding. Natural selection operates to remove potentially harmful inbreeding depression, providing ΔF is not too large so that the natural selection pressure can accumulate (chapter 8).
4. The current inbreeding coefficient of a population will depend on *both* (1) how many generations back the base generation is and (2) how rapidly the inbreeding accumulates, measured by ΔF.

Item (4) indicates that since the choice of base generation is usually an arbitrary one and independent of the biological process of the inbreeding, then the important parameter for management is ΔF. This perspective is supported by item (3).

Before looking at genetic theory, it is perhaps worthwhile examining some of the more obvious consequences of inbreeding in a population. The progress of inbreeding, in particular when ΔF is high, is associated with inbreeding depression in which traits show a steady decline in performance as inbreeding progresses. These traits are often those associated with fitness, such as reproductive ability, and survival to breeding maturity, but many other physical traits such as growth rate or mature size will also show depression. A summary of the impact of depression is given by Wiener *et al.* (1994). This depression is linked to non-additive gene actions, which underlie the observation of hybrid vigour, and it is the disappearance of heterozygosity in inbreeding (chapter 3) that is responsible for depression since it reduces the opportunities for hybrid vigour to be expressed. Traits associated with fitness are widely considered to be the most sensitive to non-additive gene actions, hence their association with inbreeding depression. Expression of inbreeding depression is often through deleterious recessive alleles, where inheriting two copies of an allele results in a lack of function, whilst inheriting 1 or 2 copies of the alternative allele results in (near) normal function. Depression is not the only consequence of inbreeding, and in breeding schemes it is often the large changes in allele frequency that result from a high ΔF that are important to manage. ΔF may be regarded as a measure of risk in a breeding scheme and Woolliams *et al.* (2002) develops this idea further.

It is valuable to consider how a layman may empirically recognise a potential inbreeding problem, or examine whether inbreeding may be a likely explanation for an observed problem in fitness. A first step may be to obtain several pedigrees back to the 8 great-grandparents and see how many and how often ancestors are common to both the maternal and paternal side for each individual. This will indicate a non-zero probability of identity by descent ($F > 0$) for any locus in the individual. Examination of the whole group of individuals may show the same great-grandparents recurring repeatedly in the pedigree. What is being noted is that a large proportion of the pathways in the genealogical tree of the population trace back to the same small handful of contemporary ancestors. There may be two reasons for this: there are only very few ancestors in this ancestral generation or there are many other ancestors but each of these others has relatively very few pathways leading down to the current population. In a heuristic way, define r_i, as the proportion of genes tracing back to each of the ancestors in the generation, and note that the sum of these proportions must be 1 because we are dividing the total gene flow between the ancestors. From the above arguments, we might be concerned if the average 'r_i' is large, or if the r_i are highly variable, or both. Consequently an empirical measure of risk would increase as the average and the variance of the r_i increase, and one function which increases in this way is the sum of squares of the r_i, denoted $\Sigma\ r_i^2$. This empirical examination is looking at the impact of just one generation, and so it is a measure of a rate of increase per generation, since the total inbreeding will depend on the accumulation of such effects over multiple generations. Therefore, a highly intuitive measure of a rate of inbreeding, ΔF, is related to the sum of squared proportions, or contributions, by a generation of ancestors to their living descendants.

2. Genetic contributions

The long-term contribution, r_i, for an ancestor i is the proportion of genes in population derived from i by descent many generations later. Note it deals only with relationship derived by descent, so for example one full-sib makes no genetic contribution to another even though they have a relationship coefficient > 0. A conceptual way of thinking about the long-term genetic contribution is that it represents the contribution of an individual's Mendelian sampling term to the long-term gene pool. This is useful because the Mendelian sampling is the unique bit of genetic variation that the individual brings to the population. The idea of the contribution of the Mendelian sampling term helps towards recognising that the gene pool of the future has contributions from all ancestors and not just the founders.

Figure 7.1 shows a small pedigree. In each generation of descendants the contributions of the ancestors to the group can be calculated. Over time, say 5 to 10 generations, these contributions converge and show no further change over time, and it is these converged contributions that are the long-term contributions. The contributions r_i will differ between the ancestors, are all ≥ 0, and will sum to 1 since a single generation of ancestors must explain the whole gene pool. The contributions of the ancestors (A ... D) to the descendants (M ... P) for the pedigree in Figure 7.1 are shown in Table 7.1: they can be calculated by working down the pedigree following the rules (a) ancestors A to D contribute 1 to themselves, 0 to the other contemporary ancestors, (b) an offspring in any generation is the average of the contributions of A to D to its sire and dam.

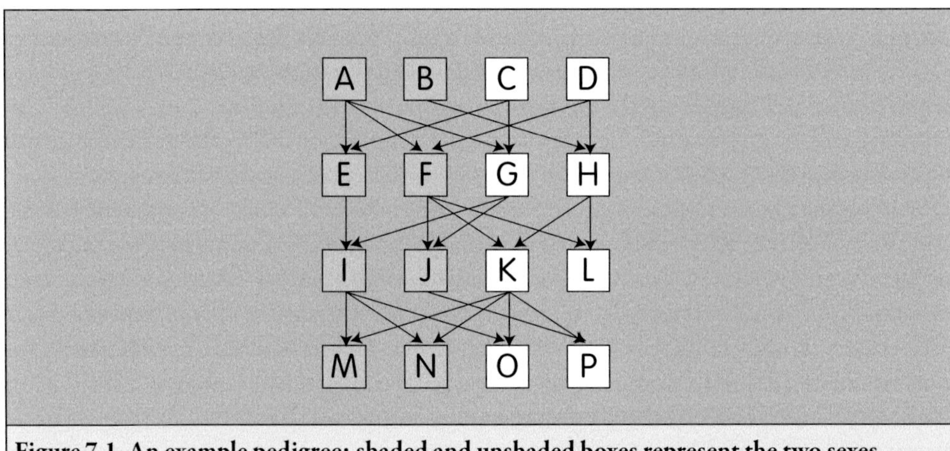

Figure 7.1. An example pedigree: shaded and unshaded boxes represent the two sexes.

Table 7.1. Contributions from ancestors A, B, C, D to M, N, O, P.

	A	B	C	D
M, N, O	⅜	⅛	¼	¼
P	⅜	⅛	⅜	⅛

An important property of the long term contributions was first proved by Wray and Thompson (1990) and extended by Woolliams and Bijma (2000) and showed the relationship of contributions to ΔF:

$$\Delta F = \tfrac{1}{4}(1-\alpha)\, \Sigma\, r_i^2 \qquad \text{(Eq. 7.1)}$$

where r_i is the long-term genetic contribution of an individual, the sum is over all individuals in the generation, and α measures the departure from fully random mating so that $\alpha > 0$ represents a preference for mating relatives and <0 avoidance of relatives. A proof is given in Box 7.2. Note the close resemblance of Equation 7.1 to the heuristic derivation of the previous section.

3. Minimising ΔF in conservation schemes

Equation 7.1 implies simple objectives for the pedigree for minimising ΔF. Minimisation requires $\tfrac{1}{4}(1-\alpha)\, \Sigma\, r_i^2$ to be minimised both within and across generations. Some well-recognised principles are then clear: (1) since the sum of long-term contributions over a generation is 1, then the sum of squares and hence ΔF is minimised if all contributors to the future gene pool make equal contributions; (2) consequently, the larger the number of individuals contributing, i.e. the number of parents per generation, then the smaller the average contribution and ΔF will be; (3) if we have two sexes, equal contributions require equal numbers of males and females since each sex must contribute ½ the genes. Note that this simple and defined target for equality of individuals' long-term genetic contributions is an important distinction from paragraph 4 where selection between families may occur.

However the simple definition of ΔF also makes clear other implications that are less widely recognised: (4) ΔF is defined in terms of the long-term contributions, *not* only the contributions into the next generation, and these contributions will develop over several generations and so minimisation of ΔF is achieved through managing how these contributions develop over time; (5) minimisation is favoured by systems that have a preference for mating relatives rather than avoidance, since loss of variation arises from segregation in heterozygotes, and their frequency is increased by matings that avoid relatives. However there are other issues associated with mating of relatives and these will be briefly discussed later.

3.1. Optimum designs

The problem of minimising ΔF when the number of males is much less than females is a frequently encountered practical problem since keeping breeding males can be a management problem. However for a fixed mating population of M breeding males and dM breeding females /generation an achievable lower bound has been established. This was described by Sanchez *et al.* (2003) after considering the minimisation of long-term contributions. Firstly consider only random mating ($\alpha \approx 0$). The study showed that: minimum $\Delta F \geq [1+2(\frac{1}{4}^{d})]/[12M] \sim 1/[12M]$ for large d, and importantly showed how this could be achieved. This is shown in Figure 2. The principles extend the initial observations of Gowe *et al.* (1959) that a sire should be replaced by a son and a daughter by a dam; and those of Wang (1997), who recognised that for d > 1 there was an imbalance whereby a dam of a selected son is favoured and should not therefore contribute a daughter, instead a different dam should contribute 2 daughters to more equally distribute the contributions. The extension of Sanchez *et al.* (2003) recognises that minimising ΔF requires management across generations and so the breeding purpose of each dam is determined from its own dam's breeding purpose, as in Figure 7.2. The distinction is that selecting without reference to previous generations allows equal initial contributions to develop into unequal long-term contributions. For example the CV of long-term contributions for the system of Wang (1997) among breeding males is $\sim 1/[2\sqrt{2}] = 0.35$ for large *d* compared to 0 for Sanchez *et al.* (2003).

These comparisons are made *without* managing mating. Wang (1997) shows that by introducing the avoidance of relatives in mating ($\alpha < 0$), ΔF can be made close to the value obtained by Sanchez *et al.* (2003) for random mating. This is effective for Wang (1997) since as a general rule $\alpha < 0$ reduces the variation of contributions about the desired expectations across generations. However Sanchez *et al.* (2003) showed that, by introducing a degree of preferential mating of relatives, consistent with the general

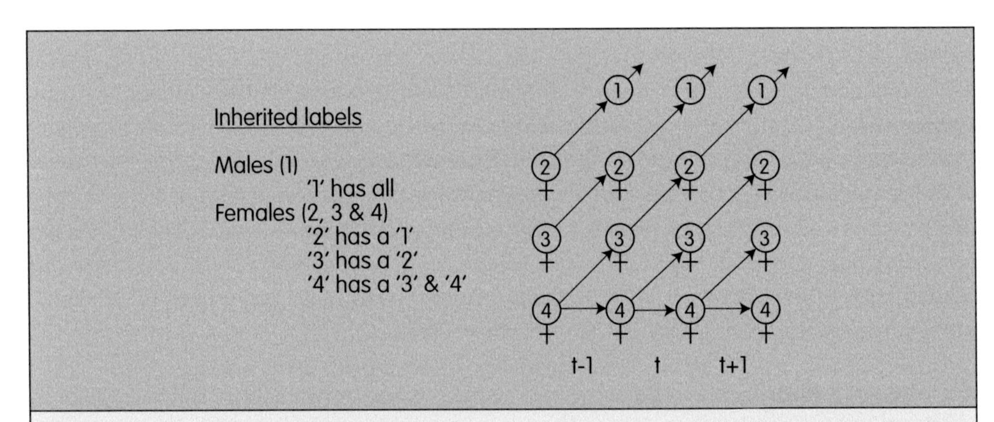

Figure 7.2. The design that achieves the lower bound for ΔF with random mating for d=3. Fifteen different individuals are shown, labelled within generations. Following Sanchez *et al.* (2003).

principles outlined at the start of this section, ΔF could be reduced further below the bound for random mating. It may be questioned whether or not the preferential mating of relatives is desirable in practice: on the positive side, (1) any degree of preferential mating is beneficial and need not be great, and (2) it is consistent with ideas of purging; on the negative side (3) there may be a greater inbreeding depression for some individuals depending on the degree of preference. Resolving such an issue will depend on the recent breeding history of the population and its potential genetic load.

3.2. Implementation

What should be done in practice? The challenges caused by implementation are exemplified by a dam that is intended to produce a son instead only having daughters or *vice versa* (not such a problem with many fish species where sex is more labile!). This requires the highly designed schemes to be robust. At present, robustness of mating schemes is still an area of debate and further work is needed. An effective robust scheme is similar to that described by Grundy *et al.* (1998) which calculates contributions after minimising the group coancestry among all parents weighted by use (not only the set of proposed matings) and utilising maximum avoidance of relatives in matings i.e. $\alpha < 0$. Here the group coancestry is $\frac{1}{2}\mathbf{c}^T\mathbf{A}\mathbf{c}$ where $\mathbf{c}$ is the vector of contributions to the next generation and $\mathbf{A}$ is the numerator relationship matrix. In a comparison Fernandez *et al.* (2003) suggest that such designs are more robust than Sanchez *et al.* (2003): however this paper ignores the principles that underpin the latter method! In particular, Sanchez *et al.* (2003) show a 'best' scheme is not necessarily based on avoidance of relatives in mating and it is reasonable to expect continuity in the underlying theory as a function of the variance of the 'noise' in producing the required offspring. While this research question is being resolved the minimisation of group coancestry is recommended and described in chapter 8.

There are some commonly done things that *should not be done* since they fail to avoid the development of unnecessary genetic bottlenecks and the consequent waste of genetic variation: (1) managing diversity by inbreeding coefficients of the parents; (2) managing diversity by the inbreeding coefficient of the offspring; (3) using what is known as the 'effective number of founders' since this only manages contributions in an arbitrary 'founding' generation and, as schemes develop, contributions from all generations need to be managed, not only founders.

Equation 7.1 also gives indications on aspects of management and the culture of management that will contribute to or hamper good genetic management. There are a number of factors that will increase ΔF as a result of introducing variation in contributions. Examples of these are:

1. Unequal mating opportunity e.g. when one sire is allocated many mates but another is allocated very few, perhaps through the unregulated use of AI for some sires, or when one sire is kept over many breeding seasons but another is culled early in its breeding life.
2. If offspring have differential survival due to differential management of families, or inherited disease or perceived 'faults'. The latter problem comes from the imposition of breed standards, usually based on exterior appearance, where offspring failing to meet standards are excluded from breeding opportunities. The dangers of this are that there is a reluctance to breed from certain individuals because their offspring have attained a bad reputation and this problem can be compounded by secrecy in the results of these examinations. The impact upon the population is to unnecessarily erode the genetic base. A more progressive strategy for both inherited disease and breed 'faults' is to use the test results to identify the degree to which the problem is genetic, and which individuals are more or less likely to carry such genes, and to develop a breeding scheme to reduce the incidence of the disease or the 'faults' in a sustainable way. This is an application of optimum contributions as developed in chapter 8.
3. Artificial selection will often introduce variation in contributions, although selection within families will have no, or only small, impact. Selection may be desirable, e.g. as mentioned above in (2), and is discussed in more detail later. Many problems are caused by fashion in favouring one sire over another e.g. according to performance in particular shows in which the animal was judged as 'best', and this can result in big demands for matings to these sires, with consequently very unequal contributions. This is a difficult social problem for many breeds, since their genetic management is shared amongst many private individuals as a hobby. However it is feasible to develop rules or quotas that retain rewards for owners of individual animals, whilst managing the impact of such activities on the long-term diversity, and these should be pursued wherever possible. More systematic genetic selection is considered in a later section and its implementation in chapter 8.

3.3. Impact of genomics

Is managing the pedigree the entire answer? In many cases yes, however Wang and Hill (2000) pointed out that effective population sizes could in theory be made infinite if selection was made actively based upon the alleles that were inherited by the offspring. This could further reduce the loss of diversity by ensuring that the replacement parents contain a balanced number of copies of each segment of DNA from each parent, although equality among segments will remain impossible if the mating ratio, $d > 1$. If $d = 1$ than this would result in no loss of diversity and an infinite effective population size. This ideal is unachievable in practical terms requiring very large families for a genome of any size. However with the prospects of dense genome wide SNP typing becoming brighter, there is a realistic prospect of selection within families to minimise the lack of balance between homologous segments of each parent in the replacement parents i.e. bringing actual contributions closer to their expectations. This concept moves us into managing diversity through evaluation as well as simply by design.

Genomic technology has developed to where we are expert at obtaining data through high throughput assays, but novices at its interpretation in relation to the range of adaptive phenotypes. For the next 5 years at least, we will have large-scale individual data based on many anonymous markers with poorly estimated effects on adaptive fitness. Additional benefits from using DNA will come solely from addressing the remaining loss of variation that lies within families: for M=10 males with large *d*, utilising 'best' management of pedigree alone gives $\Delta F=1/120$ (from Sanchez *et al.*, 2003), worse management can lead to ΔF over 5-fold greater. This general perspective on the value of primarily managing pedigree is supported by the simulations of Fernandez *et al.* (2004).

4. Selection

In species with large families the management of the pedigree to minimise ΔF can be combined with appropriate selection within families, so that some genetic gain is made. However in many commercial breeding schemes this rate of gain (ΔG) will be insufficient and alternative approaches are required that manage diversity in the presence of some degree of selection between families. In the management of diversity for such schemes, both selection and mating can play a role, however it is useful to separate these processes since whilst the principles of managing the selection are well understood, those underpinning mating are less advanced.

As stated previously the objective for pure conservation is clear: in each generation every individual should have a long-term contribution equal to those of its contemporaries. However by definition any degree of selection between families will give contributions that vary and in effective breeding schemes, this variation will be related to the Mendelian sampling term (denoted a) of the individuals. This is evident from the equation for ΔG analogous to Equation 7.1, showing that it is equal to the sum of cross products of long-term contributions and Mendelian sampling terms (Woolliams and Thompson, 1994; Woolliams *et al.*, 1999):

$$\Delta G = \Sigma \text{ ra} \qquad \text{(Eq. 7.2)}$$

Why cross products with Mendelian terms and not breeding values (denoted A)? This is because the Mendelian terms are the unique contributions of individuals, whereas

Box 7.1. Mendelian sampling terms.

For all autosomal DNA, half the genes come from the sire and half the genes come from the dam, and moreover the half that passed from each parent to the offspring are chosen at random. Therefore the expected breeding value of the offspring (A_{off}) is the average of the breeding values of its sire (A_{sire}) and dam (A_{dam}), i.e. $E[A_{off}] = ½ A_{sire} + ½ A_{dam}$, where E[ ] denotes an expectation. Expressing this as a linear regression gives $A_{off} = ½ A_{sire} + ½ A_{dam} + a$, where a is the deviation of the offspring from the average of its parents, and is called the Mendelian sampling term, with $E[a] = 0$. We can also calculate the variance of a by considering the variance of both sides of the formula for A_{off}, with the result that $Var[a] = ½(1- \alpha)\sigma_A^2$ where α is the deviation from random mating and σ_A^2 is the genetic variance in the base population prior to selection. Therefore, for random mating, the Mendelian sampling term makes up ½ the genetic variation in the base. The Mendelian sampling term arises because the actual alleles passed by each parent will vary from offspring to offspring, due to sampling among the two alleles it carries at each locus. The term is important because it makes each individual unique, not just the average of its parents, and is the source of genetic variance within families, making full-sibs different from each other.

the breeding value is an aggregation of the individual's Mendelian term and those of its ancestors, so substituting A for a in Equation 7.2 would result in double counting.

Grundy *et al.* (1998) predict that as a consequence of Equations 7.1 and 7.2, breeding schemes optimised to maximise gain for the same rate of inbreeding should allocate long-term contributions of individuals in relation to their estimated Mendelian sampling term, a prediction confirmed by Avendaño *et al.* (2004), as described below. Consequently the target contribution will change over time partly because estimates of genetic merit change, with errors reducing in magnitude as more information becomes available over time. This is not the only source of uncertainty in desired contribution: even if the breeding value of all individuals is always known with full accuracy, the desired contribution of an individual parent will change as the genetic values of the offspring become known, since their contributions cannot be determined independently without changing the long-term contribution of the parent (remember $r_{parent} = ½ \Sigma r_{offspring}$).

4.1. Optimum contributions: The problem

This is commonly referenced as Meuwissen (1997), although similar approaches were previously published by other authors (see Woolliams *et al.* (2002) for a more detailed history). The approach solves the problem of managing diversity in the course of selection by finding the solution to the following: maximise $\mathbf{c}^T\mathbf{g}$, subject to five constraints: $½\mathbf{c}^T\mathbf{A}\mathbf{c} \leq F^*$, $\mathbf{c}^T\mathbf{s} = ½$, $\mathbf{c}^T\mathbf{d} = ½$, $h \leq \mathbf{c}$ and $\mathbf{c} \leq m$, where $\mathbf{c}$ is a vector of candidate contributions to the next generation, $\mathbf{s}$ and $\mathbf{d}$ are indicator vectors for males and females respectively, m and h are upper and lower bounds to contributions respectively. The constraint $\mathbf{c} \leq$ m is eliminated if no candidate has restriction on maximum contribution. If no candidate has restriction on minimum contribution h $\leq \mathbf{c}$ becomes $0 \leq \mathbf{c}$. F* is determined from the group coancestry of the current generation of parents and the desired ΔF.

4.2. Optimum contributions: The design

There is an implicit design underlying the use of optimum contributions (see Grundy *et al.*, 1998; Avendaño *et al.*, 2004). First, in the absence of restrictions and constraints to the contrary, optimum contributions will treat males and females similarly, with equal expected numbers of parents with the same distribution of contributions in relation to estimated genetic merit. Second, the distribution of contributions has a form shown in Figure 7.3: a threshold linear relationship with estimated Mendelian sampling term, with the variance about the regression, tightly controlled (in contrast to truncation selection where this variance increases with the square of the mean).

Given the finding that the estimated Mendelian sampling term is the selective advantage (Avendaño *et al.*, 2004) it is tempting *but very mistaken* to interpret this as a form of within-family selection. What is occurring is that in each generation, from the earliest possible opportunity differential contributions are being made in relation to the best estimate of the Mendelian sampling term available at the time, so that at all stages a minimum of selection intensity is wasted between families. For example if one individual is predicted as meriting a greater long-term contribution than another why give them equal mating proportions in the initial round of selection? As the optimum contributions algorithm becomes more and more restricted by practical constraints,

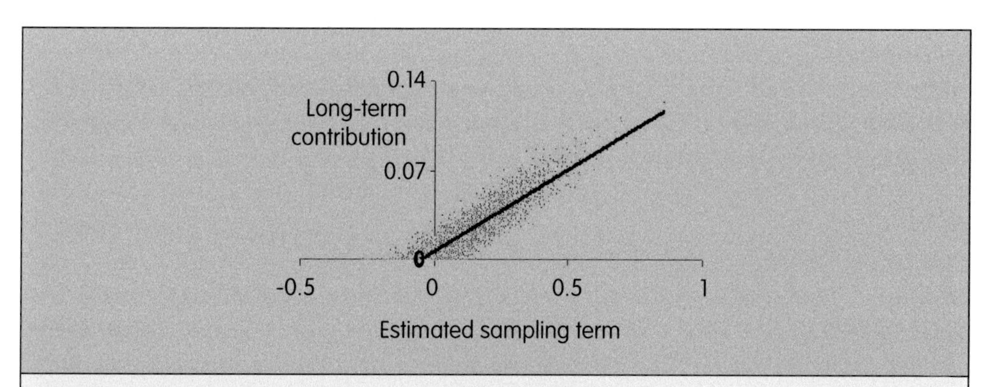

Figure 7.3. The distribution of long-term contribution when gain is maximised but diversity is managed in relation to estimated Mendelian sampling terms.

e.g. on reproductive capacity, this design will become more and more obscured in practice although the principles will remain. It is important to realise that using the optimum contribution algorithm to guide selection will *maximise* the gain made given the current rate of inbreeding that the scheme has i.e. breeding companies cannot lose by implementing the selection algorithm, and by definition *if it is not implemented the scheme is sub-optimal for gain*! The implementation of this method is described in chapter 8.

5. Predicting ΔF

An important part of designing breeding schemes is the ability to predict the impact of events on ΔF. The relationship between ΔF and long-term contributions shown in Equation 7.1 is limited for prediction since it is a function of *observed* contributions and so offers no predictions for the future, for example what happens if we select more intensely. The problem of predicting ΔF when selection in each generation is unrelated to pedigree, such as random selection, is straightforward and has long been solved. However when selection is on a trait subject to any form of inheritance the problem is more complex because each generation of selection cannot be considered independent of the previous generations. This is because the selective advantages that help an individual to be selected as a parent are passed in part to the offspring; therefore a genetically better parent is likely to have more offspring selected, and each selected offspring is more likely to produce selected grand-offspring since the grand-offspring inherit in part, but to a lesser degree, the advantage of the grandparent. Thus the selective advantage of a parent influences its long-term contribution over subsequent generations but with diminishing effect.

This problem is overcome by predicting the expected long-term contribution of an individual conditional upon its selective advantage. When selection is based on phenotype (termed mass selection) with simple inheritance, the only selective advantage for an individual i is its breeding value A_i, and $\mu_i = E[r_i \text{ given } A_i]$ or $E[r_i \mid A_i]$. Woolliams and Bijma (2000) show that for random mating ($\alpha=0$), with Poisson litter sizes, Equation 7.1 can be replaced by:

$$\Delta F = ½ \Sigma E[\mu_i^2] \quad \text{(Eq. 7.3)}$$

i.e. the observed contributions can be replaced by expectations providing the coefficient ¼ is replaced ½. For mass selection accurate predictions for ΔF can be obtained by assuming a linear model for μ_i, and the derivations of the coefficients in this model are described in Box 7.3, although these are simplified by assuming single-sexed diploids. Woolliams *et al.* (1999) show how μ_i can be derived for two sexes, overlapping

Box 7.2. The relationship between ΔF and contributions.

The following will assume for simplicity a single population of diploids with random mating. Consider a base population of N individuals at time t=0, with 2N alleles considered by convention to be neutral and distinct. Let Q be one such allele, then the allele frequency at time 0 is $f_Q(0) = (2N)^{-1}$. Define the additive trait for individual i by $p_Q(i) = 0$ when i has no Q alleles, ½ when heterozygous for Q, and 1 when homozygous for Q. In the base population, one individual i has $p_Q(i) = ½$, whilst the remainder are all zero. In all subsequent generations the frequency of Q can be decomposed into the sum:

$$f_Q(t) = \Sigma_{\text{base individuals } j}\, r_j(0,t)A_Q(j) + \Sigma_{\text{generation } u = 1 \ldots t}\, \Sigma_{\text{individuals } j}\, r_j(u,t)a_Q(j)$$

where $A_Q(j)$ is the breeding value for j for trait p_Q (½ or 0) and $a_Q(j)$ is the Mendelian sampling term for j for trait p_Q. Because this is a unique allele in the base its contribution to inbreeding at time t for random mating, $F_Q(t) = E[f_Q(t)^2]$.
$F_Q(t)$ has a very simple form, since all the cross product terms have an expectation of 0, so:

$$F_Q(t) = E\,[\,\Sigma_{\text{base individuals } j}\, r_j(0,t)^2A_Q(j)^2\,] + E[\,\Sigma_{\text{generation } u = 1 \ldots t}\, \Sigma_{\text{individuals } j}\, r_j(u,t)^2a_Q(j)^2]$$

Since the allele is neutral there is no covariance between the r_j^2 and the $A_Q(j)^2$ or $a_Q(j)^2$, and $E[A_Q(j)^2] = (4N)^{-1}$. Further note that since the base is arbitrary in a scheme of constant structure subject to constant selection pressures, then $E[\Sigma_{\text{generation}}\, r_j(u,t)^2]$ is a constant, say $E[\Sigma r_j^2]$ for all generation from 0 up to the last few where long-term contributions have yet to occur: as will be seen later this will not affect the proof and so this lack of convergence will be ignored to simplify terms. The calculation of Mendelian sampling variances will not be described, but at time t, $E[a_Q(j)^2] \approx (8N)^{-1}(1\text{-}\Delta F)^t$ with the approximation made here ignoring second order terms. Therefore:

$$F_Q(t) = E[\Sigma r_j^2](4N)^{-1} + E[\Sigma r_j^2]\,(8N)^{-1}(\Sigma_{\text{generation 1 to } t}\,(1\text{-}\Delta F)^u)$$

Now note that the inbreeding coefficient at time t is $F(t) = \Sigma_{\text{base alleles}}\, F(t)$. Since Q was an arbitrary choice F(t) is $2NF_Q(t) = ¼\, E[\Sigma r_j^2]\,(\,1 + \Sigma_{0 \text{ to } t}\,(1\text{-}\Delta F)^u)$. Let t go to infinity then F(t)=1, and the sum $\Sigma_{0 \text{ to } t}\,(1\text{-}\Delta F)^u$ is ΔF^{-1}, so after arranging terms $\Delta F = ¼\, E[\Sigma r_j^2]\,(1+\Delta F) \approx ¼\, E[\Sigma r_j^2]$. So ΔF is a direct function of $¼\, E[\Sigma r_j^2]$, and the error in $\Delta F = ¼\, E[\Sigma r_j^2]$ is O(ΔF) which is small for all practical purposes. This proof can be extended to two sexes with the same result (Woolliams and Bijma, 2000) and to non-random mating where the factor becomes $¼(1\text{-}\alpha)$, where further α relates to the departure from Hardy-Weinberg equilibrium. Note the ¼ arises from the terms describing Mendelian sampling variance, since it describes the increment in inbreeding due to the contributions uniquely attributable to each individual.

generations, and a number of inheritance models, e.g. imprinting, and selection indices. Ronnegard and Woolliams (2003) developed models with maternal effects. Equation 7.3 is then used to develop predictions of ΔF for index selection (Woolliams and Bijma, 2000), overlapping generations and mass selection (Bijma *et al.*, 2000) and truncation selection with BLUP (Bijma and Woolliams, 2000). These predictions can be obtained from software such as SelAction (Rutten *et al.*, 2002).

Some typical results are shown in Figure 7.4, where comparisons are made of the same breeding scheme structure either using random selection (hence no gain), mass selection or truncation selection with BLUP estimated breeding values. It is clear that mass selection increases ΔF, but to a much lesser degree than truncation selection on BLUP. However there is a clear difference in the relationship to heritability: for mass selection, ΔF increases as h^2 for the trait selected increases from 0, reaching an approximate plateau when h^2 lies between 0.4 and 0.7 before reducing again; whilst for truncation on BLUP, ΔF decreases steadily, converging with mass selection as h^2 tends to 1 (since additional family information is of no value when the breeding value is

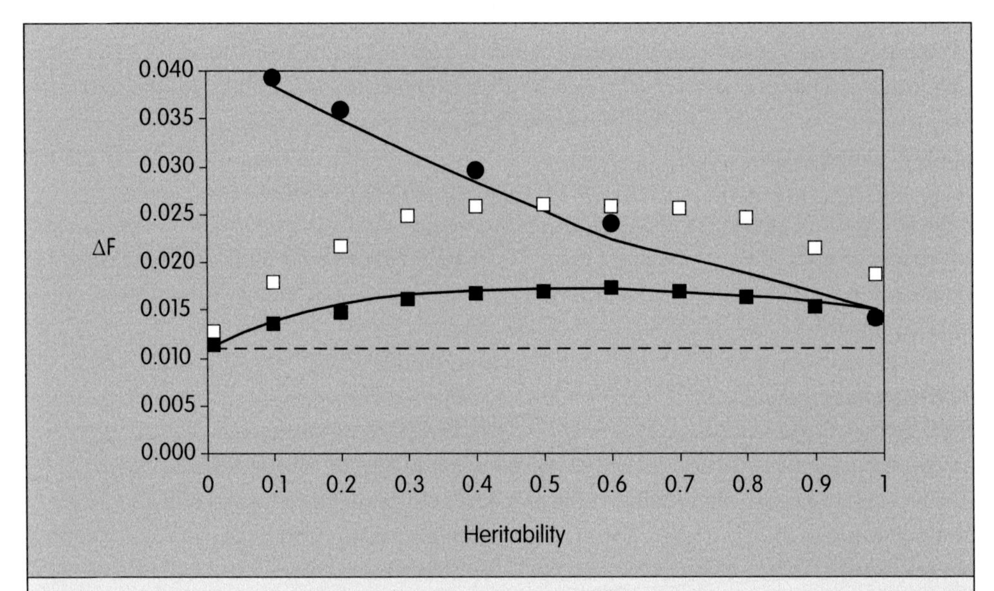

Figure 7.4. Relation of predicted (lines) and simulated (symbols) rates of inbreeding ΔF with heritability (h^2) for populations with discrete generations, with 20 sires and 20 dams and different numbers of offspring per dam (n_o, assumed fixed within a population, ½n_o of each sex): ---, random selection, n_o=8; solid squares, mass selection, n_o=8; solid circles, truncation on BLUP, n_o = 8; open squares, mass selection, n_o = 32. Based upon Bijma *et al.* (2000) and Bijma and Woolliams (2000).

Box 7.3. Expected contributions in a population with one sex.

The results on predicting gene flow and ΔF are easiest to follow in the case of a single sexed population with random mating and mass selection, where a population of T individuals are created by N selected parents and the trait selected for has heritability h^2 with breeding value for individual i denoted A_i. In mass selection the breeding value is a selective advantage that is inherited in part by the offspring. Let r_i be the long-term contribution of an ancestor i, then (ignoring the case of selfing to simplify):

$$r_i = \frac{1}{2} \Sigma_{j \text{ offspring}} r_j$$

To calculate the expected gene flow from an individual, it is necessary to calculate $\mu_i = E[r_i \mid A_i]$ and here a linear model is fitted so $\mu_i = \alpha+\beta(A_i-A_{bar(i)})$. To calculate this expectation it is simplest to consider first $E[r_i \mid A_i,n_i] = \frac{1}{2} n_i E[r_j \mid A_i]$ where n_i is the number selected from the progeny of i. The transfer of selective advantage across generations can be modelled by $(A_j-A_{bar(j)})= \pi(A_i-A_{bar(i)})$ where $A_{bar(.)}$ is the mean breeding value of the whole selected group in a generation, so that $E[r_j \mid A_i] = \alpha+\beta\pi(A_i-A_{bar(i)})$; this assumes a near-equilibrium over generations that will occur after a small number of generations. Furthermore superior ancestors will have more offspring selected, so although the average number selected per parent of a diploid species will be 2, it is modelled better by the linear approximation $2(1+\lambda(A_i-A_{bar(i)}))$. Therefore again assuming the near equilibrium $\alpha+\beta(A_i-A_{bar(i)}) = \frac{1}{2}.2.(1+\lambda(A_i-A_{bar(i)}))(\alpha+\beta\pi (A_i-A_{bar(i)})$, which allows expression of β in terms of α by equating terms in $(A_i-A_{bar(i)})$ i.e. $\beta=(1-\pi)^{-1}\lambda\alpha$. It is easily seen that on average the selected parents must have equal contributions so $\alpha=N^{-1}$; standard selection theory gives $\lambda = i\sigma_P^{-1}$ where i is the intensity of selection and σ_P is the standard deviation; also $\pi = \frac{1}{2}(1-kh^2)$ where k is the variance reduction coefficient; and consequently $\beta = 2iN^{-1}(1+kh^2)^{-1}$. Note that the value of the selective advantage is reduced by more than half in selection (since $\pi < \frac{1}{2}$) because of the increased competitiveness of the other parents. Therefore $\mu_i = N^{-1}[1+2i(1+kh^2)^{-1}\sigma_P^{-1}(A_i-A_{bar(i)})]$

From this expression of μ_i, Woolliams and Bijma (2000) show that $\Delta F = \frac{1}{2} E[\mu_i^2]$ assuming Poisson litter sizes. Therefore for the single sex population in mass selection,

$$\Delta F = (2N)^{-1}[1+4i^2(1+kh^2)^{-2}h^2(1-kh^2)]$$

ignoring terms of $O(N^{-2})$, since after selection $Var(A_i-A_{bar(i)}) = h^2\sigma_P^2(1-kh^2)$. The power of this result is that it requires only the mean conditional on the selective advantages to be modelled, which can be done for a wide class of genetic structures using the methods of Woolliams *et al.* (1999). These were developed into predictive formulae for mass selection by Bijma *et al.* (2000), and for truncation selection on BLUP by Bijma and Woolliams (2000).

measured precisely by phenotype). The graph also shows how ΔF increases with family size, since larger family sizes permit greater intensity of selection: note that only mass selection is shown in Figure 7.4, but for the same family size ΔF with truncation on BLUP will remain much greater than for mass selection until h^2 approaches 1. Note also, a point not shown in Figure 7.4, that the relationship between ΔF and ΔG is non-linear in these truncation selection schemes: for a given h^2, moving from random selection to mass selection achieves genetic gain but increases ΔF, but moving from mass selection to truncation on BLUP increases gain relatively little but ΔF dramatically.

6. Guidelines for best practice

Breeding schemes may be broadly classified into two groups: (a) those with sophisticated extensive pedigree available and where BLUP is used for evaluations; and (b) schemes which are less sophisticated or are limited in their scope to accumulate full pedigrees on offspring e.g. aquaculture of some fish species. The former schemes can have high ΔF if the estimates of breeding value are used naively, since their additional accuracy comes primarily through the use of information on relatives. As a consequence there is an increased chance of co-selection of relatives such as full-sibs, creating inappropriate variance in the long-term contributions of the parents. However, using selection algorithms such as the optimisation of contributions described in paragraph 4. allows ΔF to be managed explicitly at sustainable levels whilst retaining the primary benefit of BLUP, namely that it provides the best estimates of breeding value. Schemes sufficiently sophisticated to use BLUP are sufficiently sophisticated to utilise optimum contributions. This may not be an option in the less sophisticated schemes (but note: optimum contributions can be utilised with breeding values estimated from phenotype alone), therefore guidance from predictions is most necessary in practice for those breeding schemes using mass selection.

FAO (1998) present simple recommendations for numbers of parents to achieve Ne = 50, i.e. $\Delta F = 0.01$, in mass selection or simpler breeding schemes with the aid of 3 scenarios: (1) selection *strictly* within families; (2) selection that is *strictly* random (a dangerous assumption); and (3) mass selection with $h^2 = 0.4$. The value of 0.4 was chosen since it is at the left edge of the plateau in the relationship between h^2 and ΔF, and would be considered as a high heritability in practice. It is always safest to plan with case (3). Recommendations for these scenarios, based on Equation 6 of Bijma *et al.* (2000), are given Table 7.2 but in a slightly different format from FAO (1998). Table 7.2 shows the *minimum* number of sires required to achieve Ne = 50 in a generation, for a wide range of mating ratios and lifetime family sizes for a female. In the table it is assumed that the number of dams per male is always 1 or more, that mating is hierarchical (a conservative assumption), and that generations are discrete. An

Table 7.2. The minimum number of sires to be used per generation to achieve an effective population size of 50 or more, for different mating ratios and expected family sizes, and assuming discrete generations. The values for mass selection further assume $h^2 = 0.4$. All assume that family sizes have a Poisson distribution prior to selection. The values for random and within family selection are independent of the expected family size. Estimates are based upon: equation 6 of Bijma *et al.* (2000) for mass selection; $(8M)^{-1}(1+d^{-1})$ for random selection, Wright(1969); $(16M)^{-1}(1½+½d^{-1})$ for within family selection, Gowe *et al.* (1959).

Mating ratio	Lifetime offspring						Random selection	Within family selection
	4	8	12	16	20	36		
5 or more	21	23	25	27	28	30	15	10
4 to 5	21	25	27	28	29	32	16	11
3 to 4	23	26	28	30	31	35	17	11
2 to 3	25	29	32	34	36	40	19	11
1 to 2	31	38	43	46	48	55	25	13

approximation for overlapping generations can be made by defining the mating ratio to be the total number of breeding dams entering the breeding system per generation (= L, see Box 7.4 for information on generation intervals) divided by the total number of breeding males entering the breeding system per generation. The value of ΔF is relatively insensitive to the mating ratio above 5, so there is little to be gained from further separating out mating ratios.

Note again the major impact that mass selection can have, and the danger of assumptions that selection is random or within families when it is not so. However this should be put into perspective: (a) mass selection will deliver much faster rates of gain than strict within-family selection, particularly when litter sizes are small; (b) mass selection has a relatively benign impact on ΔF when compared to simple truncation selection in similar-sized schemes using breeding values estimated from BLUP or classical sib-indices as shown in Figure 7.4.

This justifies the recommendations:

- if using mass selection, and optimum contributions is not an option, use Table 7.2 to guide the size of scheme necessary to achieve $Ne \geq 50$;
- if the breeding scheme is sufficiently sophisticated to use evaluation methods such as BLUP, then it is both desirable and achievable to implement optimum contribution methodology, and to ensure $Ne \geq 50$.

Box 7.4. How long is a generation?

In population genetics the generation interval (L) is the length of time taken to renew the gene pool. Over the long term it is natural to consider only those ancestors destined to contribute to the gene pool in the long term. This is calculated from long-term contributions (Woolliams *et al.*, 1999): $L = (\Sigma\ r_i)^{-1}$ where the sum of long-term contributions is taken over those born over the time unit (e.g. years).
In practice, since long-term contributions are not available, it is approximated as the average age of the parents when their replacements are born. For the purpose of analysis of breeding schemes this is best estimated for each of the separate flows of genes in the population e.g. male parents to breed male replacements, male parents to breed female replacements etc. However very useful information can be obtained by simply calculating the average age of male parents at the birth of their offspring (L_m) and similarly for female parents (L_f) and calculating $L = ½ (L_m + L_f)$. The average is taken because males and females each contribute half the gene pool. This expression is used in chapter 8.
The approximation above can often overestimate the generation interval in the long-term, particularly with mass selection with a fixed age structure in the breeding herd, since if a scheme is making genetic progress, a younger individual in the breeding herd has a selective advantage over the older individuals, making its offspring more likely to be selected. Therefore the replacements are more likely to be born to parents earlier in their breeding lifetime rather than later.

References

Avendaño, S., J.A. Woolliams and B. Villanueva, 2004. Mendelian sampling terms as a selective advantage in optimum breeding schemes with restrictions on the rate of inbreeding. Genetical Research 83: 55-64.

Bijma, P. and J.A. Woolliams, 2000. Prediction of rates of inbreeding in populations selected on best linear unbiased prediction of breeding value. Genetics 156: 361-373.

Bijma, P., J.A.M. van Arendonk and J.A. Woolliams, 2000. A general procedure for predicting rates of inbreeding in populations undergoing mass selection. Genetics 154: 1865-1877.

FAO. 1998. Secondary Guidelines for the National Farm Animal Genetic Resources Management Plans: Management of Small Populations at Risk. FAO, Rome, Italy.

Fernandez J., M.A.Toro and A. Caballero, 2003. Fixed contributions designs vs. minimization of global coancestry to control inbreeding in small populations. Genetics 165: 885-894.

Fernandez J., M.A. Toro and A. Caballero, 2004. Managing individuals' contributions to maximize the allelic diversity maintained in small, conserved populations. Conservation Biology 18: 1358-1367.

Gowe, R.S., A. Roberston and B.D.H. Latter, 1959. Environment and poultry breeding problems. 5. The design of poultry control strains. Poultry Science 38: 462-471.

Grundy, B., B. Villanueva and J.A. Woolliams, 1998. Dynamic selection procedures for constrained inbreeding and their consequences for pedigree development. Genetical Research 72: 159-168.

Meuwissen, T.H.E., 1997. Maximising the response of selection with a predefined rate of inbreeding. Journal of Animal Science 75: 934-940.

Ronnegard, L. and J.A. Woolliams, 2003. Predicted rates of inbreeding with additive maternal effects. Genetical Research 82: 67-77.

Rutten, M.J.M., P. Bijma, J.A. Woolliams and J.A.M.van Arendonk, 2002. SelAction: Software to predict selection response and rate of inbreeding in livestock breeding programs. Journal of Heredity 93: 456-458..

Sanchez, L., P. Bijma and J.A.Woolliams, 2003. Minimizing inbreeding by managing genetic contributions across generations. Genetics 164: 1589-1595.

Wang, J.L., 1997. More efficient breeding systems for controlling inbreeding and effective size in animal populations. Heredity 79: 591-599.

Wang, J.L. and W.G. Hill, 2000. Marker-assisted selection to increase effective population size by reducing Mendelian segregation variance. Genetics 154: 475-489.

Wiener, G., G.J. Lee and J.A. Woolliams, 1994. Consequences of inbreeding for financial returns from sheep. Animal Production 59: 245-249.

Woolliams, J.A, and P. Bijma, 2000. Predicting rates of inbreeding: in populations undergoing selection. Genetics 154: 1851-1864.

Woolliams, J.A. and R. Thompson, 1994. A theory of genetic contributions. Proceedings of the 5th World Congress Applied to Livestock Genetics 19: 127-134.

Woolliams, J.A., P. Bijma and B.Villanueva, 1999. Expected genetic contributions and their impact on gene flow and genetic gain. Genetics 153: 1009-1020.

Woolliams, J.A., R. Pong-Wong and B. Villanueva, 2002. Strategic optimisation of short and long-term gain and inbreeding in MAS and non-MAS schemes. Proceedings of the 7th World Congress on Genetics Applied to Livestock Production 33: 155-162.

Wray, N.R. and R. Thompson, 1990. Prediction of rates of inbreeding in selected populations. Genetical Research 55: 41-54.

Wright, S. 1969. Evolution and the genetics of populations. Vol. 2. The Theory of Gene Frequencies. University of Chicago, Chicago.

Chapter 8. Operation of conservation schemes

Theo Meuwissen
Department of Animal and Aquacultural Sciences, Norwegian University of Life Sciences, Box 1432, Ås, Norway

Questions that will be answered in this chapter:

- *How to set up live conservation schemes:*
  - *What is the required effective population size?*
  - *What are the principles of managing genetic variation?*
  - *How to select and mate to minimise inbreeding?*
  - *How to select and mate in small populations?*
  - *How to select across several populations?*
  - *How to perform marker-assisted-selection?*
- *How to set up cryo-conservation schemes?*
- *How to integrate live and cryo-conservation?*
  - *How to set up cryo-backup live conservation?*
  - *How to set up cryo-aided live conservation?*

Summary

This chapter describes the minimum effective population size needed for a population to survive in the longer term, and this depends among other factors on past effective size. The general recommendation is 50. Selection and mating methods that manage genetic diversity whilst achieving genetic improvement for the breeding goals are described together with similar methods that minimise the inbreeding. In addition, the situation is considered where we want to genetically improve the population in a particular direction. The option to temporarily use some genetics from a related breed to alleviate the inbreeding is treated as well. The selection methods are extended to the use of the information of genetic markers and known genes. The importance of monitoring in a breeding scheme is stressed. Prolonging the generation interval can be a very important method to increase effective population size and reduce genetic drift. Combinations of live and cryoconservation schemes are considered where cryoconservation is used as a backup for the live scheme and where it is used to increase effective population size of small populations.

1. What issues are important?

The operational issues of conservation schemes depend on which kind of conservation plan was chosen in chapter 2 and 6. The main distinction is between pure live conservation schemes, pure cryoconservation schemes, and a combination between live and cryoconservation schemes. Within the live conservation schemes, one can distinguish *in situ* and *ex situ* live conservation, but this distinction is not very relevant for this chapter because the issues that are important for *in situ* schemes are also important to *ex situ* live schemes.

The issues that are important for live schemes are:
- the effective population size at which the breed is maintained;
- the selection of animals within the breed;
- the mating of the selected animals;
- the genetic improvement that needs to be achieved;
- the monitoring of traits and pedigree.

These issues will be addressed in paragraph 2. With respect to the issue of the selection of animals, note that some selection is possible, when sires produce more than one son and dams more than one daughter and the population is not increasing in size, but the selection may well be *at random* (instead of for a trait).

The most relevant operational issue for cryoconservation schemes is the replenishment, i.e., replacement, of retrievals from the genome bank, since retrievals will deplete the genetic materials in the genome bank. The operation of the genome bank is described in paragraph 3. The aim of the bank is to conserve genotypes rather than alleles, because, generally, we want to conserve combinations of alleles, i.e. the genotype, that leads to a characteristic of a breed instead of a particular allele.

With respect to the combination of live conservation schemes and cryo-conservation schemes there are two aims:
1. A live conservation scheme is conducted while cryo-conservation serves as a back up in case the live population runs into genetic problems (inbreeding; genetic diseases; loss of genetic characteristics; physical loss of a large part of the population). If old 'back-up' genetic material is retained, the genome bank will keep track of the full history of the evolution of the population.
2. Cryo-conservation can be actively used to increase the effective population size of a small live breed, and reduce genetic drift.

The former aim is an extension to a pure live conservation scheme, and can reduce risks substantially in these schemes. The latter aim implies a judicious use of cryoconservation, which will be given special attention in this chapter.

The combination of live and cryoconservation can result in very potent conservation strategies because:

- It can achieve all the objectives for conservation, namely opportunities to meet future market demands, insurance against future changes in production circumstances, insurance against the loss of resources with a high strategic value, opportunities for research, present socio-economic value, cultural and historic reasons and ecological value (chapter 1).
- It can reduce the genetic drift substantially, and resembles in that respect a pure cryo-conservation scheme where genetic drift is very small.
- In the combination of an *in situ* live and cryo-conservation scheme the population will still evolve and adapt to the environmental circumstances.

In the combination of an *in situ* live and cryoconservation scheme we have to find a balance between the latter two aspects: reducing genetic drift by using old, perhaps very old, cryoconserved stocks and promoting genetic adaptations by using little cryoconserved stocks. How to find this balance will be described in paragraph 4.

2. Live conservation schemes

The effectiveness of live conservation schemes depends on the effective population size and the management of the genetic variance through an effective selection and mating of the animals.

2.1. The effective population size

From conservation biology theory, effective population sizes should exceed 500 animals; otherwise the accumulation of slightly deleterious mutations will deem the population to extinction (Lynch *et al.*, 1995). However, there has arisen a controversy over the mutation rate that was assumed. Lynch *et al.* assumed a mutation rate of 0.5 mutations per genome per generation (mutation model A), while new estimation methods resulted in much smaller estimates of 0.03 mutations per genome per generation (Garcia-Dorado *et al.*, 1998; Caballero and Garcia-Dorado, 2003) (mutation model B).

Also the mean selective advantages of mutations differed between the mutation models A and B and are estimated at 0.02 and 0.2, respectively, in *Drosophila* (Caballero and Garcia-Dorado, 2003). Hence, under the model B, mutations are rarer but have larger effects than under model A. Note that deleterious mutations with large effects are less likely to drift to high frequencies in the population, because natural selection will prevent this. Hence, both these changes in the mutation model B result in much smaller effective population sizes that are needed to prevent a build up of mutational load. The critical effective population size reduced to 50 under mutation Model B (Garcia-Dorado, 2003).

Meuwissen and Woolliams (1994) balanced the drift of current deleterious mutations against natural selection, which purges deleterious mutants. Hence, they avoided the use of inaccurate estimates of mutation rates. Consequently, their results apply only in the evolutionary short term (say up to 20 generations), where the effects of a build up mutational load are still negligible. In practice, these results may be more relevant than the evolutionary long term results since 20 generations comprises 20 – 100 years for most species, and may be even longer in cryo aided live conservation plan (see paragraph 4). If at some point in the distant future a considerable load of mutations accumulated, still an action plan needs to be devised to (fine-scale) map the genes with the deleterious effects, and use marker assisted selection to get rid of the harmful alleles. When varying the assumptions of their model, the authors concluded that the critical effective size, i.e., the size below which the fitness of the population steadily decreases, is between 50 and 100 animals, provided selection was not for traits that were negatively correlated to fitness.

Although more research is needed on this subject we will assume here that the minimum effective size of a live population is 50 animals per generation, which yields a rate of inbreeding of 1% per generation. Note however that the actual size of the population may have to be substantially larger than 50 because of unequal numbers of males and females or selection (see Table 7.2 for a comparison of actual to effective sizes under mass selection). Also, a larger population can maintain more deleterious mutations because genetic drift is lower, and thus a population that was until recently kept at a much higher effective size than 50, should not be suddenly reduced to a size of 50. The latter would increase the drift and many deleterious alleles might drift to high frequencies. The recommendation of an effective size of 50 assumes however that the effective size is closely monitored and managed, and action plans are undertaken as soon as effective size drops below 50.

2.1.1. Effective population sizes when generations overlap

Most livestock populations have overlapping generations, i.e., the parents of new born animals are not all strictly of the same age, e.g. some may be 2 and other 3 years old. We will assume here that drift, and thus effective size, is to be controlled per generation.

The alternative is to control drift per year. The difference between controlling drift per generation or per year becomes clear when we consider a cryoconservation scheme, where a population of 50 animals is re-established after 100 years of cryostorage. In this scheme, the drift per year is small but that per generation is as large as usual for a population size of 50. Also, the cryoconserved population did not achieve any genetic adaptation during the last 100 years. Hence, the minimisation of drift per year can result in no genetic adaptation, whereas the minimisation of drift per generation allows for the fact that the population has to evolve. Because genetic adaptation is one of the main aims of an *in situ* live conservation plan, the genetic drift should be minimised per generation, i.e., the effective size should be maximised per generation. For *ex situ* live conservation plans also some genetic adaptation will be desirable in most cases and the same argument holds.

Since the effective size *per generation* is relevant in populations with overlapping generations, also the numbers of sires and dams selected has to be expressed per generation. The number sires used per generation is the number of newly introduced sires per year times the average generation interval (averaged over sires and dams). Similarly, the number of dams used per generation can be calculated.

2.1.2. Prolonged generation intervals

The above implies that, if the numbers of sires and dams selected per year are relatively constant, an increase of the generation interval will result in an increased number of sires and dams selected per generation, and thus to an increased effective population size. *Thus, prolonging the generation interval can be a very important method to increase effective population size and reduce genetic drift.* However, as mentioned before, we also want to turn over generations in order to achieve natural and artificial selection response, and therefore a judicious use of long generation intervals is recommended.

2.1.3. Monitoring of effective population size

When pedigree is recorded, the coefficient of inbreeding can be calculated for every animal (Falconer and Mackay, 1996). Hence, the increase of the average inbreeding coefficient can be calculated per year. However, the increase of inbreeding is due to the increase of the average kinship in the population, which makes the mating of completely unrelated animals impossible. Thus, the future average inbreeding is described by the current coefficient of kinship. Hence, more up to date results are obtained, by calculating the average kinship across all pairs of animals as:

$K_a = ¼K_a(m) + ¼K_a(f) + ½K_a(mf)$,

where $K_a(m)$, $K_a(f)$, and $K_a(mf)$ are the average kinships between all pairs of males, females, and male-female pairs, respectively (excluding pairs of an animal with itself). Thus, the yearly increase of the average kinship coefficient can be calculated $\Delta K(yr)$, and is expected to equal the future rate of inbreeding (strictly this assumes that we are dealing with one population, not several populations that do not mix). If further, the average age of the sires and dams at birth of their offspring is recorded, i.e., the generation interval (L) is recorded, we can calculate the increase of the kinship per generation as $\Delta K(gen) = L\,\Delta K(yr)$. The effective population size is now $Ne = 1/(2\Delta K(gen))$ animals per generation. This ΔK will equal the observed future ΔF. It is important to monitor effective population sizes, because they can be smaller than expected due to any effect that increases the variance of the family size of the animals (e.g., selection, unequal survival rates). *Hence, it is very important to keep records of the pedigree in live population schemes, i.e., to store the sire and dam identification number of every animal in a data bank.* If pedigree is recorded, inbreeding and kinship can be calculated, together with their rates of increase per generation or per year.

2.2. Principles of the management of genetic variation

The management of inbreeding (and thus genetic variation) in a breeding scheme relies very much on the control of the average relationship of the selected parents. In a population, the average relationship of the parents including self-relationships equals the average relationship of their offspring excluding self-relationship, because:

$$A_{ij} = .25^*[A_{SiSj} + A_{SiDj} + A_{DiSj} + A_{DiDj}] \qquad \text{(Eq. 8.1)}$$

where A_{ij} is additive genetic relationship between offspring i and j ($i \neq j$), S_i (S_j) is sire of i (j), D_i (D_j) is dam of i (j).

Equation 8.1 shows that:

1. The relationship among two random individuals i,j equals the average of the relationships of their parents, which may be generalised to the average relationship among offspring equals the average relationship of their parents weighted by how many offspring the parents contributed.
2. Since the sire of i may also be the sire of j ($S_i = S_j$), the relationships of the parents with itself should be included when calculating the average relationship among parents. Thus, in the following, average relationship of parents implicitly means including relationships with itself, and average relationship of offspring implicitly means excluding relationships with itself.

The offspring i and j may be mated (if mating is at random) and obtain themselves an offspring, say x, then the inbreeding coefficient of x is:

$F_x = ½ A_{ij} = K_{ij}$

where K_{ij} = the kinship coefficient of i and j.

Theory shows that although the concepts of relationship and kinship are different, they are simply related as $K_{ij} = ½ A_{ij}$. The Equation $F_x = K_{ij}$ implies that, if mating is at random, the average inbreeding of the grand-offspring of F_x equals the average kinship of the parents of K_{ij} that we are currently selecting. If mating is not at random, for instance if we mate the least related animals with each other, the average inbreeding of the grand offspring will be lower than the average kinships of the current parents. However, the averaging involved in Equation 8.1, which occurs every generation, make that the differences in relationships between the current parents average out quickly, and after a few generations the average relationship due to the current parents is the same across the entire population, and every possible mating will translate this average relationship into inbreeding. Thus, maximum avoidance of inbreeding mating can delay inbreeding due to the average kinship of the current parents, but after a few generations, the inbreeding will occur anyway. For instance, if generation 0 consists of unrelated animals, generation 1 contains some full-and half sib relationships, and the first inbred animals are expected in generation 2. If we avoid the mating of full and half sibs, generation 2 animals will still be non-inbred, and generation 3 will show the first inbred animals, but the rate of increase of the inbreeding from generation 3 onwards will be the same as that for a breeding scheme without avoidance of sib-matings (where this rate of inbreeding started already in generation 2).

An exception is if we split the population in separate lines, in which case the between line kinship will not be translated in inbreeding, because animals are not mated across lines (however, due to the smaller population sizes within the lines, inbreeding will be increased). For an interbreeding population, this implies that the rate of inbreeding equals the increase of the average kinship of the parents, and different mating systems can delay the inbreeding but *the rate is unaffected*. Although, in many selection methods, the mating strategy can set up matings that affect the average kinship of the selected parents and thereby indirectly affect the rate of inbreeding. However, it seems easier and more effective to directly control the average kinships of the selected parents in the selection step. Optimum Contribution (OC) selection is such a selection method (Meuwissen, 1997; Grundy *et al.*, 1998). Mating methods that set up favourable matings for OC selection are minimum coancestry mating (which is maximum avoidance of inbreeding mating) and factorial mating (mate (as much as possible) every sire with every dam; Woolliams, 1989), and a combination of these two mating methods (Sonesson and Meuwissen, 2000).

2.2.1. Mate selection

A strategy that optimises the selection and mating in one step is called mate selection (Kinghorn *et al.*, 2002; Fernandez *et al.*, 2001). Mate selection usually tries to maximise a selection criterion that contains genetic gain, average relationship of the selected parents, and average inbreeding of the offspring, where the weighings of these three components need to be defined by the user. Since it is difficult to obtain such weights, and because in conservation schemes we would like to have control over the rate of inbreeding, instead of it being the result of a complicated weighing and optimisation process, we will focus on OC selection here, which directly controls the rate of inbreeding.

2.2.2. Optimum Contribution selection and Factorial-Minimal-Coancestry mating

Generally we advise the following selection and mating strategy:

- Step 1: Use Optimal Contribution selection to determine how many offspring each animal should get. The selection criterion is either to minimise the average kinship among the selected parents or to maximise genetic improvement whilst limiting the average kinship among the selected parents (paragraph 2.3).
- Step 2: Use Factorial-Minimum-Coancestry mating to decide on who is mated to whom, where factorial implies that each mating pair obtains only 1 offspring (or at least as few as possible offspring) (see Box 8.1).

Step 1 determines the rate of inbreeding; Step 2 tries to set up matings which facilitate the OC selection in the next generation. Step 2 also avoids the mating of close relatives, which is important to avoid the occurrence of the occasional highly inbred offspring. Since, Step 1 determined the future effective size of the population, it is much more important than Step 2 in the management of genetic variation.

2.2.3. Random - Non-sib matings

The most straightforward method to mate the individuals is assigning the matings at random, e.g., when a sire should get 10 offspring from 10 dams, the dams are sampled at random to the sire. However, some of the matings will be between close relatives, i.e., full-sib and half-sib matings. This should be avoided since the offspring from these matings will be highly inbred and will thus show increased inbreeding depression.

Box 8.1. Factorial-Minimum-Coancestry mating.

Suppose the OC selection of paragraph 2.3 resulted in a contribution of c_i for animal i:

Optimal number of offspring from sire i = $m_i = 2Nc_i$
Optimal number of offspring from dam j = $f_j = 2Nc_j$

where the sum of the m_i (as well as f_i) equals N = the total number of offspring.

A (natural) mating between a sire and a dam yields n offspring, where we make n as small as practically possible to achieve 'Factorial mating' as closely as possible. Then the number of matings needed with sire i is $M_i = m_i/n$, and with dam j is $F_j = f_j/n$, where some rounding[1] of these figures will be required to get the desired total number of matings ($N_{mat} = N/n$).

The following steps may than be used to assign the minimum coancestry matings:
Step 1: Set up (at random) a mating table, where Mat(i,1) denotes the sire of the i-th mating (i=1,.., N_{mat}) and Mat(i,2) the dam of the i-th mating.
Step 2: Use the following algorithm to perform minimum coancestry mating, where K[x,y] denotes the coancestry (kinship) between animal x and y:

```
For k=1,2,3,..etc.
    Pick at random a mating x (1≤x≤N_mat)
    Pick at random another mating y (1≤y≤N_mat; x≠y)
    Set REL_current ⇐ K[Mat(x,1),Mat(x,2)] + K[Mat(y,1),Mat(y,2)]
    Set REL_test ⇐K[Mat(x,1),Mat(y,2)] + K[Mat(y,1),Mat(x,2)]
    Count the number of succesfull swaps: S_k ⇐ S_k-1
    If (REL_test < REL_current) then
        Swap the dams of the matings: Mat(x,2) ⇐ Mat(y,2) & Mat(y,2) ⇐ Mat(x,2)
        Set S_k ⇐ S_k+1
    End if
    If S_k== S_k-100: Finish (there were no more successful swaps)
End for loop
```

When the algorithm has finished, the Mat-table contains the minimum coancestry matings.

[1]For example if $m_i/N = 4.459$, we round this figure to $M_i = 4$. However after such rounding the sum of the M_i (ΣM_i) may not equal the desired number of matings, N_{mat}. If $\Sigma M_i < N_{mat}$, we give one extra mating to the sires that were rounded down the most. Similarly, if $\Sigma M_i > N_{mat}$ we subtract one mating from the sires that were rounded up the most.

2.3. Selection and mating to minimise inbreeding

2.3.1. Optimal Contribution Selection to minimise inbreeding

A robust recommendation for pure conservation programmes is the minimisation of the average kinship, because it minimises inbreeding and maximises allelic diversity (chapter 7). Although the strategies of chapter 7, that minimise sums of squared contributions, will minimise inbreeding, they may sometimes be difficult to apply in practice, e.g. due to practical reproductive limitations. In such situations and also when the live population goes through a recent severe bottle neck, the family structure can be (very) unbalanced. Minimum kinship selection will attempt correct the unbalanced contributions of historical families and minimises the genetic drift. However, whenever possible, in the long term the conservation strategies of chapter 7 should be re-instated. With minimum kinship selection, a group of parents is selected that minimises:

$$K_a = \Sigma_i \Sigma_j c_i c_j K_{ij},$$

where Σ_i (Σ_j) denotes summation over all selection candidates; K_a is the average kinship of the selected animals; K_{ij} is coefficient of kinship between animals i and j; c_i is the contribution of animal i to the next generation, i.e., $c_i = ½ n_i / N$, n_i is number of offspring from animal i, and N is total number of offspring (the ½ is because a sire (dam) contributes only half of its genes to the offspring). Also, $n_i = 2Nc_i$ gives the number of offspring a parent should have given its optimal contribution c_i. The optimal contribution c_i that minimises K_a is given in Box 8.2. When the family structure is balanced, minimum kinship selection will result in within family selection.

The genetic drift can be controlled at the DNA level by calculating the kinship conditional on genetic marker information. The markers are used to improve the coefficient of kinship between the animals in the sense that the kinship is assessed at the DNA level, whereas kinships that are calculated from the pedigree alone are expected coefficients of kinship of the DNA segments. However, Fernandez *et al.* (2005) found no advantage using this approach compared to using the pedigree alone.

2.3.2. Mating

Mating is by Factorial-Minimum-Coancestry mating (see Box 8.1) to decide on who is mated to whom, where factorial implies that each mating pair obtains only 1 offspring (or at least as few as possible offspring). This mating strategy will to some extend equalise the relationships between the offspring, such that in the next OC selection round, the selection of one offspring does not preclude another due to a high relationship.

Box 8.2. Optimum Contribution selection to minimise inbreeding.

It is useful to rewrite the minimum kinship selection problem, as described in the text, in matrix notation:

$$K_a = c'Kc,$$

Where K is (q*q) matrix of coefficients of kinships (q = number of selection candidates); and c is the vector of contributions. It can be shown that K_a is minimised when the contributions are:

$$c = ½ K^{-1} Q (Q' K^{-1} Q)^{-1} 1,$$

where 1 is a column vector of ones; Q is a (q*2) incidence matrix of the sex of the candidates where the first column contains a one for male and a zero for female candidates, and the second column contains a one for female and a zero for male candidates. The contributions of the male and those of the female candidates will sum to ½.

2.4. Selection and mating in small populations

The conserved live population may be well above the minimum effective size of 50 animals (see paragraph 2.1) and some improvement of the genetic adaptations of the breed may be desired as described in chapter 2. In this case two selection methods will be suggested:

- Phenotypic selection, i.e., select for own performance records of the animals.
- Optimal contribution selection (Meuwissen, 1997; Grundy *et al.*, 1998).
- Some form of Marker Assisted Selection (MAS).

The first selection method is very easy to implement, whereas optimal contribution selection is a rather high-tech method. *Note that if using BLUP (Best Linear Unbiased Prediction) for estimating breeding values, the optimal contribution method should always be used. Otherwise the effective population size can be severely reduced below the actual size and thus increase genetic drift (without the user being aware of it).*

2.4.1. Phenotypic selection

Truncation selection for phenotypic values is the simplest method of selection, but, at equal rates of inbreeding, it can outperform BLUP-selection (Quinton *et al.*, 1991). It is very easily implemented when the traits can be measured for both sexes. For example, simply select the animals with the highest growth rate. If the animals are kept in different herds, which hampers a direct comparison of animals across herds, selection can be for the standardised deviation of the animals from the herd mean (or herd-year-season mean).

When the trait is only recorded on one of the sexes, e.g., litter size, selection could be at random in the unrecorded sex. Alternatively, a number of female offspring could be obtained from the male selection candidates and selection could be for the phenotypic mean of the female offspring or the dam of the males. The latter requires however, that the population is of a quite large size.

Often selection will be for more than one trait, i.e., several traits need to be improved. As before phenotypic selection will only be for the own performances of the animals, but the own performances have to be combined into a selection index such that the population mean will change into the right direction. This involves three steps:

1. Determine the optimal direction of the selection, i.e. picture the animal that is optimally adapted to its environment (and market niche). The picture of this optimal animal should not be too optimistic such that it can be reached within a reasonable time horizon.
2. Obtain a desired gains selection index to select the animals in the optimal direction using only own performance records (see Cameron, 1997). See Box 8.3 for a brief description of the desired gains selection index.
3. Calculate the selection response that will be achieved within the time horizon using Cameron (1997). If the size of the selection response deviates substantially from the original goal of step 1, the goal of step 1 should be made more realistic and steps 2 and 3 should be repeated.

From step 2 a selection index can be calculated for every animal by weighing the own performances of the traits by the index weights. Selection proceeds as with single trait selection, but with the individual trait replaced by the selection index.

The above desired gains index avoids determining economic values for every trait, i.e. they are implicitly determined by the desired gains (Cameron, 1997). This seems useful, because the calculation of economic weights can be very complicated for traits in which local breeds often excel: fertility, disease resistance, longevity and quality of special products. Selection for over-simplified breeding goals can make the characteristics of the local breed equal to those of the introduced breed, which has usually been very intensely selected for a simple breeding goal. *In situ*ations where the calculation of economic weights is rather straightforward, the traditional optimal selection indices should be used (Cameron, 1997). Both desired gains and optimal selection indices require knowledge about genetic (co)variances amongst the traits. These are not always available for small breeds, but perhaps estimates from other related breeds can be used.

If the latter is not possible, some 'guestimates' of weights could be used and adjusted as the realised response is not in the desired direction.

It should be kept in mind that phenotypic selection will reduce effective population sizes below those for random selection, i.e., the effective size will be smaller (see Table 7.2). The reduction in effective size increases with the heritability of the trait and the intensity of selection, but a reduction of >30% should be expected.

2.4.2. Optimal contribution selection

Optimal contribution selection maximises the genetic level of selected parents, while controlling the increase of the average kinship in the population. Since the average kinship of the parents equals the inbreeding of their grand-offspring (see paragraph 2.2.), controlling this average kinship implicitly controls the rate of inbreeding.

We will describe optimal contribution selection in its simplest form. The genetic merit of the selected parents weighed by their contributions is:

$G = \Sigma_i c_i EBV_i$,

Box 8.3. Desired Gains Selection Index.

Let the x_i denote the change that is needed for trait i to move from the current population mean to the desired optimal population mean. Further let x denote a vector of these changes. The genetic variance-covariance matrix of the traits is denoted by G. It is assumed that all traits that are to be improved are also measured; otherwise the calculation of the index weights is more complicated. The optimal desired gains selection index weights can now be obtained from the vector:

$b = G^{-1} x$.

The vector of selection responses of the traits over the time horizon, T (in yrs), is:

$$\Delta \mathbf{G}_T = \frac{\mathbf{x}\, i\, T}{L \sqrt{\mathbf{b'Pb}}}$$

where: *i* is intensity of selection (see Falconer and Mackay, 1996); L is the generation interval; and P is the phenotypic variance-covariance matrix of the traits.

where Σ_i denotes summation over all selection candidates, EBV_i is the BLUP breeding value estimate; c_i is the contribution of the i-th selection candidate (as defined in paragraph 8.2.3). Note that c_i is zero for non-selected candidates. The average kinship of the selected parents is:

$$K_a = \Sigma_i \Sigma_j c_i c_j K_{ij},$$

which should increase by no more than ΔF per generation, where ΔF is the desired rate of inbreeding. Thus, we restrict the average kinship to:

$$K_r = 1 - (1-\Delta F_d)^t$$

where t is the generation number, and ΔF_d is the desired rate of inbreeding.

Meuwissen (1997) described an algorithm that optimised the contribution of each candidate, c_i, (and thus the number of offspring that each candidate should get), such that the genetic merit of the parents, G, is maximised, and such that the average kinship does not exceed K_r.

In a simulation study, this selection method realised the desired levels of inbreeding and yielded about 30% more genetic improvement than selection for BLUP breeding value estimates (at the same rate of inbreeding).

Optimal contribution selection has been extended to breeding schemes with overlapping generations by Meuwissen and Sonesson (1998) and Grundy *et al.* (2000).

Although it is more difficult to implement, optimal contribution selection is definitely favoured over selection for BLUP breeding value estimates. This is because optimal contribution selection actively controls the rate of inbreeding, whereas BLUP selection can result in rates of inbreeding that are much higher than expected based on the number of selected sires and dams.

2.4.3. Mating

Mating is by Factorial-Minimum-Coancestry mating (see Box 8.1) to decide on who is mated to whom, where factorial implies that each mating pair obtains only 1 offspring (or at least as few as possible offspring). This mating strategy will to some extend equalise the relationships between the offspring, such that, in the next round of selection, the OC selection is as little as possible hindered by the restriction on the average kinship, i.e. when two candidates have a high EBV, they can both be selected since their relationship

Box 8.4. Optimal contribution selection for genetic improvement.

In optimum contribution selection we want to maximise the genetic level of the selected parents, G, which is calculated as their EBVs weighed by their contributions, **c**, where the contributions are to be optimised. We however restrict the average kinship of the selected parents to a predefined value K_r (see paragraph 8.2.3), i.e.

Maximise: $G = c' \times EBV$

Under the restrictions: $K_r = c' \times K \times c$

$\frac{1}{2} = Q' \times c$

for the unknown c, where Q is a (nx2) matrix indicating the sex of the animals (Q(i,1) = 1 if candidate i is male; otherwise Q(i,2)=1 and Q(i,1)=0), and the last restriction makes that the contributions of all the males (females) sum up to ½. Using the lagrangian method for restricted optimisation, the optimum solution is obtained:

$$c = A^{-1}(EBV - Q\lambda)/(2\lambda_0)$$

where λ and λ_0 are lagrangian multipliers which enforce the restrictions (Meuwissen, 1997).

is similar to that between any other pair of candidates. For a more detailed description on the mechanics of mating structures see Sorensen *et al.* (2005).

2.5. Selection across breeds

In some situations it is not possible to maintain a breed as a closed breeding population, because either the numbers are too small to avoid *excessive* amounts of inbreeding, or the numbers are too small to create a (internationally) competitive breeding scheme, or some characteristics from another breed are highly needed.

2.5.1. Numbers too small to avoid excessive amounts of inbreeding

In this situation, some parents from another, preferably, related breed can be used temporarily until the number of animals builds up again, and to avoid the negative effects from the genetic bottle neck that resulted from the too small population size. A related breed is favoured here, because otherwise too much between breed diversity is lost due to changing the genetics of the endangered breed towards that of an unrelated breed. It should be noted that the use of parents from another breed should be a temporary situation, because otherwise the genes from the foreign breed will (slowly) replace those in the current population.

It is also possible to merge two too small breeds into one, in order to rescue the 'meta'-population. Again, the two breeds should be highly related. If the relationship between the breeds is similar to that within the breeds, in fact very little diversity will be lost. The MVT diversity criterion (chapter 6) will quantify the loss of diversity by merging the breeds. The extinction probabilities method of chapter 6 can be used to assess whether, in the long term, more diversity will be maintained by merging the breeds than by not merging the breeds.

2.5.2. Numbers too small for competitive breeding scheme

In the situation where the breed is decreasing in size due the competition of foreign/ other breeding schemes, the following options can be used:
1. introgress some desirable characteristic(s) of the foreign breed into the local breed;
2. merge with other threatened local breeds and set up a competitive breeding scheme for the meta-population.

1. Introgression: Introgression can be based on phenotypic information, where some animals are crossed with the foreign breed and the crossbred population is continuously backcrossed to the local breed, whilst the desirable characteristic is maintained in the population by selection. Alternatively, the genes underlying the desirable characteristic are identified and mapped by a QTL mapping experiment (see paragraph 2.6). And subsequently introgress the region(s) containing the desirable QTL allele(s) from the foreign breed in the local breed using continuous backcrossing with the local breed, and use molecular genetic markers to maintain the desirable QTL regions in the back crossed population (chapter 4; Visscher *et al.* 1996).

2. Merge with other threatened local breeds: As in the case with too small numbers to avoid excessive amounts of inbreeding, the local merging breeds are ideally highly related and the loss of diversity by merging can be quantified by the MVT criterion (chapter 6) and the extinction probabilities method of chapter 6. An important issue is whether the merging breeds (can) have the same breeding goals or not. If they have different breeding goals, and the correlation between the breeding goals is between 0.7 - 0.8, there is virtually no extra genetic improvement in economic value to be expected, but still by keeping the populations separate the genetic responses would be more into the desired directions. If the correlation is below 0.7, there may well be a loss of genetic improvement due to merging the breeds.

2.6. Marker Assisted Selection

In recent years, selection using genetic markers is receiving more and more attention, especially in cattle. Here, the steps involved are described and the benefits discussed.

2.6.1. QTL detection

QTL detection can be based on F2 crosses of populations or can be performed in outbreeding populations. The principles of QTL detection from F2 crosses have been dealt with in chapter 4. The F2 cross design is becoming less and less popular probably because the basic assumption of the F2 - QTL mapping design, namely that the parental populations are inbred, is not met in animal populations. Also a detected QTL may not be segregating within the parental populations and thus can not be used directly. However, the power of the F2-design does not breakdown completely when the assumption of inbred lines is not true. The outbreeding design is mainly based on linkage analysis within big (halfsib) families. Such large half sib families are probably not available in populations of limited size. Combined linkage and linkage disequilibrium analysis (Meuwissen *et al.*, 2002) relies less on big family sizes (Lee and Van der Werf, 2004), but requires denser marker maps. A detailed description of QTL mapping methods and designs can be found in Weller (2001).

2.6.2. Linkage Analysis - Marker Assisted Selection (LA-MAS)

After a QTL has been detected and more than 5% of the total genetic variance is explained by the QTL, the QTL can be used for LA-MAS. LA-MAS requires no population wide linkage disequilibrium between the QTL and the marker(s), and thus can be applied when the markers are still quite far away from the QTL (10 – 20 cM), or the QTL is not very precisely mapped. LA-MAS starts by calculating marker assisted BLUP breeding value estimates using the Fernando and Grossman's (1989) model. Next selection is based on these marker assisted BLUP breeding value estimates instead of the conventional BLUP breeding values. OC selection is still required to control the build up of kinship and inbreeding in the population. It may be noted however, that OC selection as presented above will control the inbreeding at a random position in the genome, not at positions close to the QTL, where the inbreeding will be substantially higher. Most MAS breeding programs do not control inbreeding near the QTL position, because this will reduce response at the QTL.

2.6.3. Linkage Disequilibrium - Marker Assisted Selection (LD-MAS)

Here, a marker has been detected that is so close to the QTL, that it can be used as a direct indicator of the QTL allele. LD – MAS relies on linkage disequilibrium between marker and the QTL, which means that the marker and the QTL are associated. Many genetic tests are based on LD markers, except those where the causative mutation has been detected (e.g. DGAT1; Grisart *et al.*, 2002). The association between the marker and the QTL is likely to differ from one population to the next, and, in fact, the association may well be opposite in different populations due to the occasional recombination between the QTL and the marker. *Thus, it is very important to (re-)estimate the association between the LD marker and the QTL in the population where it is to be used.* Also, the association between the marker and the QTL can change over time (generations), and needs to be re-estimated every 2 or 3 generations (depending on the intensity of MAS and effective population size).

2.6.4. Gene Assisted Selection (GAS)

In the case of GAS, the causative mutation underlying a genetic difference between animals has been found (such as: DGAT1; Grisart *et al.*, 2002), and selection can be directed at the positive allele. The effect of a causative mutation is much less likely to differ from one population to the next, and is expected to be relatively constant over time. Differences of effects of causative mutations can still occur due to interactions with the background genes in the population. E.g. some background genes may increase the effect others may reduce it.

2.6.5. Genomic Selection

Genomic selection is the selection for thousands of LD markers simultaneously using a high density marker map (Meuwissen *et al.*, 2001). With genomic selection, the effect of a single marker is not proven significant, but it is assumed that the sum of all the estimated effects will pick up all genetic variance in the genome. The estimation errors on all these thousands of effects are expected to average out, and be close to zero. Thus, genomic selection makes selection for the total genetic variance possible, whereas the fore mentioned MAS strategies concentrated on one or few QTL, and thus concentrated on a limited part of the total genetic variation. Control of the genetic variation by OC selection, either by assessing the kinships using markers, using pedigree or using both, will be essential, because otherwise the strong selection for particular chromosomal regions can markedly reduce the genetic variability in large parts of the genome.

2.6.6. Marker Assisted Conservation

*In situ*ations with very dense markers and small genome size, even the within family drift (the drift due to Mendelian sampling) can be stopped by marker assisted selection of a set of animals where the contribution of the paternal allele and maternal allele are equal for every parent and every position in the genome (Wang and Hill, 2000). This ultimate and high-tech control of genetic drift may be useful in some situations where the size of the population has been reduced to very few animals (e.g. a couple of males and 10 females).

3. Cryo-conservation schemes

Users of cryoconserved genetic stocks should replenish the genetic material as far as possible. To what extent genetic material can be replenished will depend on the kind of use of the genetic material (FAO, 1998):

- Embryos used for re-establishing the breed can be replenished by storing embryos from the re-established breed into the genome bank. Special care has to be taken with respect to the maintenance of genetic variation: minimum kinship or within family selection should be used to get the population that results from the thawed embryos out of its bottle neck, and the same selection should be used to select the embryos that replenish the cryo-bank.
- Semen used to re-establish the breed should also be replenished by the re-established breed. Again the maintenance of genetic variation is important. A problem is here that the stocks used for replenishing are not 100% pure bred, even after repeated back crossing with the conserved breed. However, the genes of the replenished semen should come for more than 90% from the conserved breed. This would be obtained by 4 generations of repeated back crossing.
- A genome bank of somatic cells can easily be replenished, since the retrieval requires thawing and further culturing of the cell line. The cultured cell line can be cryoconserved again.
- Semen used to create a 'synthetic' breed from several founder breeds can not be replenished by semen from the original breed.
- Semen used for crossbreeding studies (including F2 QTL mapping designs) can also not be replenished.
- Semen used for introgression of genes from the genome bank breed into another breed can not be replenished.

Fortunately, the latter three uses of the genome bank require rather small amounts of semen, but still the bank has to be replenished. Therefore, these uses of semen should provide funds for a full replenishment of the breed in the genome bank. The latter involves re-establishing the breed from embryos (or semen) to replenish the bank. Also, the information that is obtained from genetic studies, crossbreeding and re-establishing the breed (and any breed comparison involved in that) are very valuable and should be entered into the data bank.

4. Integrating live and cryoconservation

4.1. General aspects

Integrated live and cryoconservation schemes resemble very much the live schemes of paragraph 2 and hence the operational issues of paragraph 2 also apply here. There are two kinds of integrated live and cryoconservation schemes:

- The cryo-back-up live scheme. A live scheme where cryostored old genetic material is used as a back up in case (genetic) problems occur.
- The cryo aided live scheme. A live scheme with prolonged generation intervals which are achieved by cryostorage.

It seems risky to run a live conservation scheme without any back up in case of emergencies. Hence, the cryo-back-up live scheme is recommended for any live scheme, especially, if the effective population size does not exceed 50. Although, an effective size of 50 is considered to be safe, still inbreeding depression can lead to poor fertility at some point in the future, or genetic defects can become frequent. Also, diseases and other catastrophes can easily wipe out or severely reduce the small or geographically isolated population, and these are good reasons for cryo-back-up regardless of effective size. Continuous cryostorage of semen and embryos seems therefore a very useful tool to minimise these risks.

The cryo-aided live scheme is useful when the (financial) resources are not sufficient to achieve the minimum effective size of 50 in the live population. With the cryoconserved stocks the generation interval can be increased until the effective size of 50 animals per generation is reached. The increase of the generation interval reduces the rate of genetic adaptation (and improvement) of the breed and should therefore be minimised.

4.2. Cryo-back-up live conservation schemes

4.2.1. Back up in case of physical disaster

In case of a catastrophe (disease, fire, etc.), we might need to re-establish (part of) the population. The risks of physical disasters are substantially reduced by keeping the population at several herds (also the males). If a large part of the population dies, the remainder of the population will go through a genetic bottleneck. Minimum kinship selection may be used to get the population out of the bottleneck. If the use of sires across the separate breeding herds is limited, the storage of semen of sires from the previous generation can alleviate the bottleneck further, and decreases the gap between the material stored and the population that is lost.

Natural disasters spread over a large region (floods, diseases, tornadoes) may kill the entire population. For these cases we have to be prepared to re-establish the breed. Re-establishing a breed is easier with stored embryos than with stored semen. Therefore some cryostorage of embryos seems to be the preferred method. The following strategy seems sufficient to achieve a reasonably recent back-up for re-establishing the breed:

1. At the beginning of the conservation scheme, cryoconserve embryos from the founder population. The numbers needed are the same as those for a pure cryoconservation scheme that aims at re-establishing a breed (chapter 2).
2. Cryoconserve semen from every sire that is used in every generation of the scheme. This will cost little effort in an AI based scheme. In case this is too costly, 50 sires can be stored per generation, where the sires are as little as possible related. This is sufficient to maintain an effective size of 50 in the stored sire population.
3. Repeat step 1 every 5 – 20 generations. The figure of 20 generations seems reasonable for a scheme without any selection, i.e., we only want to capture new genetic adaptations of the breed. The figure of 5 generations seems reasonable in a scheme with strong selection, where we do not want to lose much genetic progress. Since the costs of cryoconservation are mainly due to getting the material and cryoconserving it, it seems a safe and relatively in-expensive strategy to maintain the earlier cryoconserved material.

In case of an emergency, we implant the most recently stored embryos into recipients and mate the offspring with the most recent semen stored in step 2. Hence, if it helps reducing the cost, we can store only female embryos. The re-established population will show a genetic lag to the deceased population of at the most 3 (=½(5+1)) or 10.5 generations, if we store embryos every 5 or 20 generations, respectively. These figures apply when the disaster occurred just before the cryoconserved stocks were due to be supplemented. Since the disaster may occur at any time since the latest collection of stocks, on average the lag will be 1.75 or 5.5 generations, respectively.

In order to avoid problems due to natural disasters, the cryo bank should be kept at a different place than the live animals and preferentially at two different places, where the second place duplicates the storage of the first place.

4.2.2. Back up in case of genetic problems

Despite an effective size of 50, a deleterious mutation can drift to a high frequency in the population. In the case of a recessive genetic defect, the frequency of the mutation is rather high before the defect is discovered in the homozygote animals that show the disease. This is because as long as the defect is at low frequency it will be mainly present in a heterozygote form in the population and the defect will not appear. However, when we discover a genetic defect, we will not know its mode of inheritance (single gene or several genes; recessive or dominant). The genetic defect can be treated like any other genetic characteristic and can be reduced in frequency by OC selection combined with BLUP estimation of breeding values, or in case that the defect is known to be due to a single gene, combined with breeding value estimates that account for the single gene nature of the defect (Sonesson *et al.*, 2003). It is important to control the inbreeding when removing genetic defects, especially in the situation where few animals are left that do not carry the disease. In such a situation, strict selection for non-diseased animals would send the population into a bottle neck. The back up storage of semen and embryos as described in the previous section may be used to increase the effective size of the population. However, any information on the carriers of mutation might avoid re-introducing it. Note also that the older animals will have the highest breeding value estimates since the population was deteriorating over time.

In the previous paragraph we suggested OC selection against a genetic defect. OC selection can also be applied to remedy genetic problems that are due to many genes. For example, poor reproductive performance of the animals can be remedied by selection for reproductive performance. Paragraph 2.3 describes how to select for any trait. If the poor reproductive performance is considered to be due to inbreeding depression, the use of old cryostored genetic stocks can be useful to reduce the levels of inbreeding in the population. Paragraph 2.2 describes how to minimise the kinship of the animals and thus the levels of inbreeding. Again the storage strategy as described in the previous section seems to produce sufficient amounts of cryostored stocks. For the purpose of reducing the kinship of the population, it is however important to retain old cryoconserved material because the older the material the lower the kinship coefficients.

4.3. Cryo aided live conservation schemes

4.3.1. Storage of embryos

In order to demonstrate the principles, we will first consider a cryo aided live scheme that has a very long generation interval of, for example, 25.5 years. The time from birth to the production of the embryos is 1 and 2 years, for sires and dams, respectively, the

embryos are frozen for 23 years, and the embryos take 1 year to develop into a new-born offspring (a total 25.5 years on average). This scheme involves in year t:
1. Thaw the embryos that were frozen 23 years ago and implant them into a recipient female.
2. Get embryos by mating the 1-year-old sires and 2-year-old dams and freeze these embryos.

It is important that the sires and dams used in step 2 are from different ages such that the generations overlap. Overlapping generations can also be obtained by using for 50% of the embryos 1-year-old sires and for 50% 2-year-old sires, and similarly for dams: 50% 1-year-old dams and 50% 2-year-old-dams. In practice, we should choose the most convenient way to achieve the overlap between the generations.

It will take some time to set up a cryo aided live scheme, because initially there will not be 23-year-old embryos available. The scheme can be set up by freezing many embryos from the founder population, such that these embryos can be used in step 1 until they are actually 23-year-old.

If step 2 involves one male and one female, i.e., every year one male and one female is raised, the scheme involves 25 males and 26 females per generation, since the male and female generation interval is 25 and 26 years, respectively. The effective size is thus approximately 51 animals per generation. Hence, in this cryo aided conservation scheme a zoo-sized population reaches an effective size of 50.

4.3.2. Storage of semen

A perhaps somewhat hypothetical scheme that uses frozen semen is a scheme that stores large quantities of semen of N founder sires in generation 0 (G0), and keeps on using this semen to fertilise live females generation after generation. In this scheme, the genes from the N G0-sires will replace all other genes in the population, and inbreeding asymptotes to 1/(2N), i.e. the rate of inbreeding goes to zero. An improvement to this scheme was suggested by Sonesson *et al.* (2002) where also the semen from N generation 1 (G1) males was stored. The semen of the G1-males was used for females from even generation numbers, and that of G0 males was used for females from odd generation numbers. In this scheme the inbreeding asymptotes to 1/(3N), and also half of the genes of founder females are conserved. Again the rate of inbreeding is zero. Further improvement was achieved by an optimal rotational crossing scheme of the sires, which reduced inbreeding to $½^N$ (Shepherd and Woolliams, 2004). Since these schemes stop the evolution of the population, they are only applicable to very limited situations, e.g. in the case of zoo-populations or populations in nature parks.

Semen based schemes with some turn over of generations and thus evolution could resemble the cryo aided live scheme with embryo storage. For example in the following scheme, 1 and 2-year-old females are inseminated with 23-year-old semen which was produced by 1-year-old males, to obtain offspring when 50% of the dams are 2 and 50% are 3-year-old. The average generation interval is 13.25 years (=½(½*1+½*23) + ½(½*2+½*23)).

This scheme involves in year t:
1. Inseminate the 1-year-old (50%) and 2-year-old (50%) females with the 23 year-old-semen.
2. Obtain semen from 1-year-old males and cryoconserve it.

The use of females from different ages is again to ensure that the generations overlap. If we raise 6 males and 6 females per year in this scheme, and also use them in step 2, the number of males and females per generation is 24*6=144 and 2.5*6=15, respectively. Assuming that random selection and mating is applied, this yields an effective size of 54 (Falconer and Mackay, 1996). It should be noted that the extension of the male generation interval alone is far less effective in increasing effective size than extending both generation intervals (see storage of embryos section).

After some number of years of extending the male generation interval, it hardly helps to extend it any further. Plotting the number of years of storage of the semen in a graph against the effective size, will show where the effect of increasing the male generation interval starts to level off. It is deleterious to the conservation scheme to extent the male generation interval beyond this number of years, because every extension of the male generation interval slows down the rate of genetic adaptation of the population.

It is very important to take great care when setting up a semen based cryoaided conservation scheme. If we would simply store semen from the founder males and use that for the first 23 years (analogous to the embryo storage case) rather than accumulating it over time, we would greatly increase the genetic contribution of the founder males over that of the founder females. This could *half* the size of the founder population. There are three methods to remedy this problem:
1. Make sure that the size of the male founder population is twice as large as initially intended.
2. Set up the semen cryo-aided scheme with embryos, as described in the embryo section and start to run the semen aided scheme when the founder population is 23 years old in the presented example. In practice this number of years may be much smaller.

3. Start breeding from the founder population using minimum kinship selection (paragraph 2.2) and store semen from the sires. Note that within the minimum kinship selection procedure also the 'stored' males are selection candidates, i.e., even dead males can be selected because their semen is stored. Again, as soon as the population is '23 years' old the original semen cryo aided scheme can be run, where '23 years' will be much shorter in most practical situations. It is not useful to keep on running the minimum kinship selection scheme, because it will try to prevent all genetic drift and thus evolution when the frozen ancestors become more and more old.

Because of the high tech selection involved in method 3, the methods 1 or 2 may be more practical. However, methods 1 and 2 require more resources in terms of embryos or founder animals, which can be more costly than the high tech selection method 3.

As mentioned before, cryo aided schemes can be very useful to increase the effective size of small populations, but care should be taken to keep on turning over the generations such that the population can evolve. The actual size of the population will often be determined by the available (financial) resources, and the generation interval follows from the actual size and the effective size that needs to be achieved. The cryostorage involved in the cryo aided schemes serves automatically as a genetic back up that is needed as described in paragraph 4.2.

References

Cameron, N.D., 1997. Selection indices and prediction of genetic merit in animal breeding. AB International, Oxon, UK.

Caballero A. and A. Garcia-Dorado, 2006. Genetic architecture of fitness traits: lessons from Drosophila Melanogaster. Proceedings 8th World Congress on Genetics Applied to Livestock Production, CD-ROM Communication No. 29-01.

Falconer, D.S. and T.F.C. Mackay, 1996. Introduction to quantitative genetics. 4th edition, Longman, Harlow, Essex, UK.

FAO, 1998. Management of Small Populations at Risk. Secondary Guidelines for Development of National Farm Animal Genetic Resources Management Plans, FAO, Rome, Italy.

Fernandez, J., M.A. Toro and A. Caballero, 2001. Practical implementation of optimal management strategies in conservation programmes: a mate selection method. Animal Biodiversity and Conservation 24: 17-24.

Fernandez, J., B. Villaneuva, R. Pong-Wong and M.A. Toro, 2005. Efficiency of the use of pedigree and molecular marker information in conservation programmes. Genetics 170: 1313-1321.

Fernando, R.L. and M. Grossman, 1989. Marker-assisted selection using best linear unbiased prediction. Genetics Selection Evolution 21: 467-477.

Garcia-Dorado, A., J.L. Monedero and C. Lopez-Fanjul, 1998. The mutation rate and the distribution of mutational effects of viability and fitness in Drosophila Melanogaster. Genetica 103: 255-265.

Garcia-Dorado, A., 2003. Tolerant versus sensitive genomes: the impact of deleterious mutation on fitness and conservation. Conservation Genetics 4: 311-324.

Grisart, B., W. Coppieters, F. Farnir, L. Karim, C. Ford, P. Berzi, N. Cambisano, M. Mni, S. Reid, P. Simon, R. Spelman, M. Georges and R. Snell, 2002. Positional Candidate Cloning of a QTL in Dairy Cattle: Identification of a Missense Mutation in the Bovine DGAT1 Gene with Major Effect on Milk Yield and Composition. Genome Research 12: 222-231.

Grundy, B., B. Villanueva and J.A. Woolliams, 1998. Dynamic selection procedures for constrained inbreeding and their consequences for pedigree development. Genetical Research 72: 159-168.

Grundy, B., B. Villanueva and J.A. Woolliams, 2000. Dynamic selection for maximising response with constrained inbreeding with overlapping generations. Animal Science 70: 373-382.

Kinghorn, B.P, S.A. Mesaros and R.D. Vagg, 2002. Dynamic tactical decision systems for animal breeding. Proceedings of the 7th World Congress on Genetics Applied to Livestock Production 33:179-186.

Lee S.H. and J.H.J. van der Werf, 2004. The efficiency of designs for fine-mapping of quantitative trait loci using combined linkage disequilibrium and linkage. Genetics Selection Evolution 36:145-61.

Lynch, M., J. Conery and R. Burger, 1995. Mutation accumulation and the extinction of small populations. American Naturalist 146: 489-518.

Meuwissen, T.H.E. and J.A. Woolliams, 1994. Effective sizes of livestock populations to prevent a decline in fitness. Theoretical and Applied Genetics 89:1019-1026.

Meuwissen, T.H.E., 1997. Maximising the response of selection with a pre-defined rate of inbreeding. Journal of Animal Science 75: 934-940.

Meuwissen, T.H.E. and A.K. Sonesson., 1998. Maximising the response of selection with a pre-defined rate of inbreeding II Overlapping generations. Journal of Animal Science 76: 2575-2583.

Meuwissen, T.H.E., B.J. Hayes and M.E. Goddard., 2001. Prediction of total genetic value using genome-wide dense marker maps. Genetics 157: 1819-1829.

Meuwissen, T.H.E., A. Karlsen, S. Lien, I. Olsaker and M.E. Goddard, 2002. Fine mapping of a Quantitative Trait Locus for twinning rate using combined Linkage and Linkage Disequilibrium mapping. Genetics 161: 373-379.

Quinton M., C. Smith and M.E. Goddard, 1991. Comparison of selection methods at the same level of inbreeding. Journal of Animal Science 70: 1060-1067.

Shepherd, R.K. and J.A. Woolliams, 2004. Minimising inbreeding in small populations by rotational mating with frozen semen. Genetical Research 84: 87-93.

Sonesson, A.K. and T.H.E. Meuwissen, 2000. Mating schemes for optimum contribution selection with constrained rates of inbreeding. Genetics Selection Evolution 32: 231-248.

Sonesson, A.K., M.E. Goddard and T.H.E. Meuwissen, 2002. The use of frozen semen to minimise inbreeding in small populations. Genetical Research 80: 27-30.

Sonesson., A.K., L.J. Janss and T.H.E. Meuwissen, 2003. Selection against genetic defects in conservation schemes while controlling inbreeding. Genetics Selection Evolution 35: 353-68.

Sorensen AC, P. Berg and J.A. Woolliams, 2005. The advantage of factorial mating under selection is uncovered by deterministically predicted rates of inbreeding. Genetics Selection Evolution 37: 57-81.

Visscher, P.M., C.S. Haley and R. Thompson, 1996. Marker-Assisted Introgression in Backcross Breeding Programs. Genetics 144: 1923-1932.

Wang, J. and W.G. Hill, 2000. Marker-assisted selection to increase effective population size by reducing Mendelian segregation variance. Genetics 154: 475-489.

Weller, J.I., 2001. Quantitative Trait Loci Analysis in Animals. CABI Publishing. London.

Woolliams, J.A., 1989. Modifications to MOET breeding schemes to improve rates of genetic progress and decrease rates of inbreeding in dairy cattle. Animal Production 49: 1-14.

Chapter 9. Practical implications of utilisation and management

Erling Fimland[1] *and Kor Oldenbroek*[2]
[1] *Nordic Gene Bank Farm Animals, P.O. Box 5003, N-1432 Ås, Norway*
[2] *Centre for Genetic Resources, the Netherlands, P.O.Box 16, 6700AA Wageningen, the Netherlands*

Questions that will be answered in this chapter:

- *Which international regulations exist?*
- *What needs to be organised to manage national farm animal genetic resources?*
- *How are farm animal genetic resources valued?*
- *How can farm animal genetic resources be managed in a sustainable way?*
- *What indicators exist for sustainable management?*
- *What future policy is needed for utilisation and management?*

Summary

The starting point in this chapter is that the countries that ratified the Convention on Biological Diversity (CBD) accepted this convention as an international law. The stage of implementation of the CBD in national regulations differs between the Contracting Parties (countries) of the CBD. The development of a strategic action plan for the utilisation and management of national farm animal genetic resources is proposed as a first step. The sustainable use of farm animal genetic resources is outlined in accordance with the CBD: "the use of the components of biological diversity in a way and at a rate that does not lead to the long-term decline of biological diversity, thereby maintaining its potential to meet the needs and aspirations of present and future generations". The values of farm animal genetic resources can be categorised as the sum of direct use values and several more hidden values to safeguard genetic diversity as a resource for future production environments and for securing all other potential values of the local breeds. These values are linked to securing future food supply, expressed as satisfaction of society needs. This implies that there should be governmental interest in the long-term effects of breeding schemes, and that this interest is expressed in a national strategic plan for sustainable use of farm animal genetic resources.

1. Existing global regulations

The United Nation Convention on Biological Diversity (CBD) was negotiated within the framework of the UN Environment Program (UNEP), and was opened for signature in 1992 (see Box. 9.1). Today the Convention has been ratified by more than 180 States and should safeguard the sustainable management of biological diversity. Sustainable use is defined in the CBD as: "the use of the components of biological diversity in a way and at a rate that does not lead to the long-term decline of biological diversity, thereby maintaining its potential to meet the needs and aspirations of present and future generations" (Art. 2). The nations that have ratified the CBD have accepted the principles of the CBD as an international law. This implies that they have accepted a responsibility for the utilisation and conservation of the national farm animal genetic resources. A logical first step for a country is developing a national strategy and an action plan for farm animal genetic resources.

The national agricultural authorities should take the lead to develop a strategy and an action plan in collaboration with the other stakeholders in the field of farm animal genetic resources. The involved parties should include, at least, breeding organisations and breed societies that have the formal right to manage the country's breeding schemes and other organisations with responsibilities for the conservation of endangered breeds. In some countries, national gene resource committees exist, while in other countries, a national bureau of agriculture or a national centre for genetic resources appointed by the agricultural ministry fulfil a central role in the implementation of the strategy into action plans.

The CBD is the only international law calling for tools that secure farm animal genetic diversity for future use. At the international level, the Agreement on Trade-Related Intellectual Property Rights (TRIPS) regulates patents or other intellectual property rights on living organisms, including farm animals (see Box 9.2). The TRIPS agreement was negotiated under the World Trade Organisation (WTO). It is a comprehensive agreement in international law, as it encompasses a variety of intellectual property rights. However, a substantial number of WTO member states still has to implement the TRIPS provisions in national law and policy, whereas the different patent laws in the member states are not harmonised and may differ in requirements and protection offered. It does require that all its members shall provide for legislation on patents (including patents on living organisms, cells and genes) regarding inventions in any field of technology (TRIPS-agreement article 27.1.), (NCM, 2003). In connection to these provisions, the TRIPS agreement offers certain exemptions, as can be seen from article 27.3: "*Members may also exclude from patentability:......(b) plants and animals other than micro-organisms, and essential biological processes for the production of plants or*

Box 9.1. The Convention on Biological Diversity (CBD) (FAO, 2004).

The Convention on Biological Diversity (CBD), though not focusing on Animal Genetic Resources(AnGR) as such, does cover all kinds of genetic resources. Article 2 of the CBD defines genetic resources as "genetic material of actual or potential value" and further defines genetic material as "any material of plant, animal, microbial or other origin containing functional units of heredity".

The three objectives of the CBD, as set out in Art. 1, are: the conservation of biological diversity, the sustainable use of components of biological diversity, and the fair and equitable sharing of the benefits arising from the utilisation of genetic resources.

Although not directly stated in the CBD, conservation of biological diversity necessarily includes conservation of animal and plant genetic resources, which are the prerequisites for food security and the improvement of agricultural productivity. The CBD states that, while nations have the sovereign right to exploit their own resources (Art. 3), they also have the duty to conserve them. The need for policy development and integration is acknowledged in the CBD, and governments are requested to develop national strategies on biodiversity (Art. 6a), and to integrate "the conservation and sustainable use of biological diversity into relevant sectored and cross-sectored plans, programmes and policies" (Art. 6b).

The benefit-sharing dimension of the third objective of the CBD, which is "the fair and equitable sharing of the benefits arising out of the utilisation of genetic resources" as stated above, includes appropriate access to genetic resources and appropriate transfer of relevant technologies, taking into account all rights to those resources and technologies, as well as funding.

With regard to access to genetic resources, Art. 15 of the CBD recognises the sovereign rights of States over their natural resources, and states that access is subject to national legislation (Art. 15.1). Access is to be granted on mutually agreed terms (Art. 15.4), therefore through bilateral agreements. This implies that both supplier and recipient of genetic material must agree on the terms and conditions of the transfer, and that, unless otherwise determined by that Party, prior informed consent of the Contracting Party providing the genetic resources applies (Art. 15.5).

The legal provisions in such a bilateral agreement can be taken to mean that the provider of genetic resources must be fully informed in advance by the access-seeking party about the objectives, as well as the economic and environmental implications of such access. The CBD foresees the necessity of legislative, administrative or policy measures to provide for fair and equitable sharing of the results of research and development and the benefits arising from the commercial and other utilisation of genetic resources with the Contracting Party providing such resources (Art. 15.7).

A benefit-sharing component is also found in Art. 8(j), which contains provisions to encourage the equitable sharing of the benefits arising from the utilisation of knowledge, innovations and practices of indigenous and local communities, embodying traditional lifestyles relevant for conservation and sustainable use of biological diversity.

Box. 9.2. The WTO Trade-Related Intellectual Property Rights Agreement (TRIPs) (FAO, 2004).

TRIPS, which has been in force since January 1995, is the broadest multilateral agreement on intellectual property that applies to: copyright and related rights, trademarks, including service marks, geographical indications, including appellations of origin, industrial design, patents, including the protection of new varieties of plants, the layout designs of integrated circuits, and undisclosed information, including trade secrets and test data.
Under Art. 27.3 of TRIPS, WTO Members must protect various forms of intellectual property, some of which are relevant to AnGR, including indications of geographical origins, trademark, trade secrets and patents. TRIPS requires Members to make patents available for any inventions, whether products or processes, in all fields of technology without discrimination, subject to the normal tests of novelty, inventiveness and industrial applicability.
There are three permissible exceptions to the basic rule on patentability. The exception relevant to AnGR is contained in Art. 27.3 (b), stating that Members may exclude "plants and animals other than micro-organisms, and essentially biological processes for the production of plants or animals other than non-biological and microbiological processes".
Most countries worldwide have explicitly excluded patents for animals. It will be many years at least before animals are treated equally with other applications in the patent system. Animal patenting may become an issue with the introduction of transgenic production animals.
Even where animals or parts thereof are deemed, in principle, patentable, a patent application may be rejected on moral or public order grounds, in accordance with Art. 27.2 of TRIPs. Nevertheless, the notions of morality and public order are quite vague and changing, and their content will depend on national perceptions by patent offices and judges. In fact, the determination of whether certain conduct may be contrary to fundamental values of a society is a matter of national public policy.

animals other than non-biological and microbiological processes. However, Members shall provide for the protection of plant varieties either by patents or by an effective sui generes system or by any combination thereof...."

It is important to recognise that plant varieties have a convention for protection of breeding activities through the Union for the Protection of New Varieties of Plants (UPOV), which is an example of a *sui generis* system that ensures property rights to the breeders, but weaker property rights than patents. Such a system does not exist for the animal breeding sector for the development of breeds or lines by breeding organisations. UPOV is a formal organisation with the task of evaluating and approving, according to defined rules, that new varieties are distinct, uniform and stable, and have improved properties compared to earlier approved varieties. Such a system for breeders' exemptions is not appropriate for breeds of farm animals, because a substantial amount of genetic diversity within a breed is a necessity for its fitness and for further selection aimed at genetic improvement (see Box. 9.3).

The CBD states as a principle that: "States have sovereign rights to exploit their own resources pursuant to their environmental policies". This principle raises very important issues for animal breeding and conservation, which should be solved in the near future:
- the ownership of genetic resources and the rights for access to these resources;
- the rights (patents) on technologies required to exploit genetic resources;
- the responsibilities for conservation and sustainable use;
- the fair and equitable benefit sharing arising from the use of genetic resources, especially those which are maintained by local communities.

A complication in the CBD framework arises when the farm animal genetic resources are maintained developed and supplied by international breeding companies, as is already the case in poultry and pigs. This development requires an international perspective on conservation. This complication requires attention from the international organisations involved (WTO, FAO and CBD-parties).

2. An example of a national action plan for utilisation and management

This section presents an outline for organising the management of farm animal genetic resources within a country. It is obvious that national traditions and rules should be taken into account. The first step in this process is the identification of all possible actors such as:
- the ministries involved in agriculture, environment and food;

Box. 9.3. The World Intellectual Property Organisation (WIPO) (FAO, 2004).

The World Intellectual Property Organisation (WIPO) is an international organisation whose mandate is to ensure that the rights of creators and owners of intellectual property are protected worldwide and that inventors and authors are thus recognised and rewarded for their creativity. WIPO's Intergovernmental Committee (IGC) on Intellectual Property and Genetic Resources, Traditional Knowledge and Folklore was established in 2000. The committee provides "a forum for international policy debate about the interplay between intellectual property and traditional knowledge, genetic resources and traditional cultural expressions (folklore)". The committee has met five times in 2003 and 2004. The current key questions are a possible International Instrument on Intellectual Property in Relation to Genetic Resources and on the Protection of Traditional Knowledge and Folklore, and a possible requirement that patent applications include a mandatory disclosure of the source of genetic material used.

- the national breeding companies and associations for reproduction technology (artificial insemination and embryo technology);
- the national centre of genetic resources or a coordinating unit for genetic resources;
- breed societies of mainstream and endangered breeds;
- research institutions working in the field of animal breeding and genetics.

In some countries, it is natural to include representatives from the public sector or consumer organisations in the preparation of a national action plan.

Fimland (2006) illustrates an example (that may apply to many countries) of a national organisation and the collaboration between the central stakeholders (Figure 9.1). More generally, this scheme may provide the elements to develop an appropriate organisation in a specific situation. Numbers attached to the arrows refer to comments given below:

1. It is assumed that the agricultural ministry has given a mandate for conservation activities with an annual budget to a coordination unit for genetic resources or a genetic resource centre. This unit/centre reports annually about the activities and the costs of the centre linked to the realised work for animal genetic resources.
2. It is assumed that the agricultural ministry has stated the directives for running a breeding scheme and performing reproduction technology activities. In the European Union, this is regulated by several directives.
3. Cryopreservation can be performed much cheaper when a close collaboration exists with a breeding organisation or a reproduction technology centre. The marginal costs of these organisations in these collaborations are much lower than the full costs of a separate conservation centre. The operation of the genetic resource centre should address the following issues:
   a. The ownership rights of the conserved material should be transferred from the donor farm to the genetic resource centre. The genetic resource centre directly owns the conserved material coming from the endangered breeds.
   b. The genetic resource centre should negotiate with the reproduction technology-centre that a certain amount of semen (see chapter 2) from each endangered breed will be stored for the long term: around 10 doses of semen per bull. In addition, a plan should indicate the use of the additional storage of semen from available bulls for minimising inbreeding in the endangered breeds.
   c. The genetic resource centre should make sure that the breeding companies of the commercial breeds store around 10 doses from each bull for long-term storage.
   d. The collection and transfer of the material should always be in agreement with the existing veterinary rules.

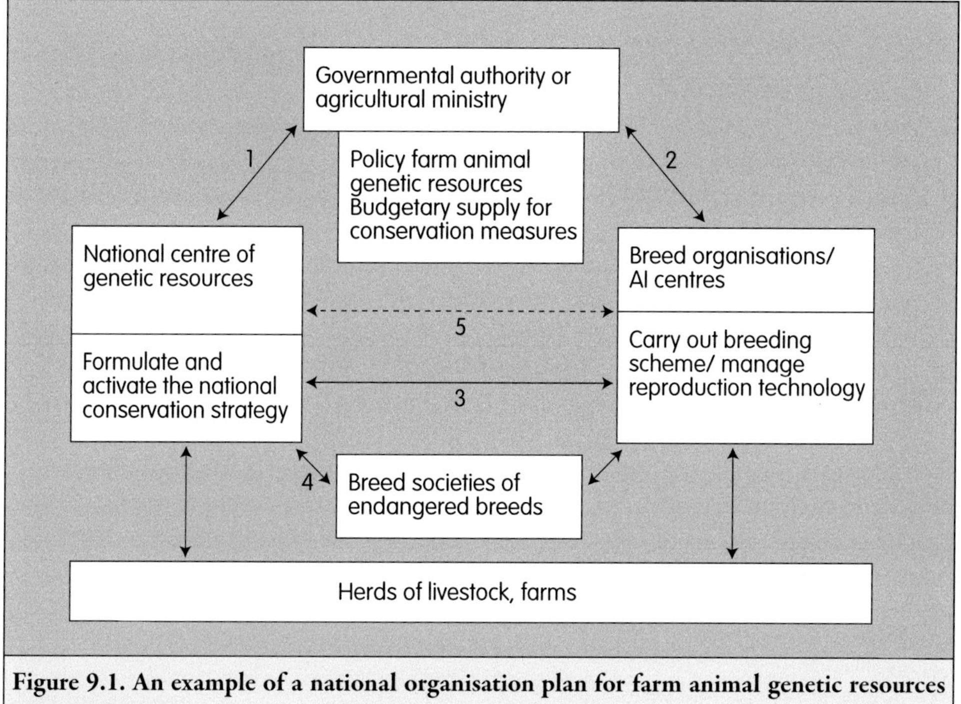

Figure 9.1. An example of a national organisation plan for farm animal genetic resources (Fimland, 2006).

4. The genetic resource centre and the breed societies of endangered breeds must collaborate on the selection of individual breeding animals to be used for conservation and it must be clear who is responsible for the different parts of the conservation activities.
5. The genetic resource centre might have a mandate to review the commercial breeding companies' work in accordance with the defined sustainability requirements.

In addition, the following may be required:
- The genetic resource centre takes care of pedigree recording, performance tests, veterinary tests and data characterisation. It also makes the relevant information available to a broader group of potential users.
- It is rational that the FAO's National Coordinator is located at the genetic resource centre or the ministry of agriculture.
- The genetic resource centre should cooperate closely with research institutions to secure the transfer of updated knowledge for sustainable management of animal genetic resources and to stimulate necessary research in this field.

Besides these operational activities, the objectives of the genetic resource centre might be to:

- secure that all stakeholders take part in policy development and are stimulated to taking responsibility;
- document the tasks and responsibilities of each of the partners and negotiate and maintain necessary agreements with the partners;
- describe the procedures relevant to the maintenance of farm animal genetic resources;
- produce an annual report describing the sustainable utilisation and management, including conservation;
- report annually on income and expenses for the main operational activities.

3. Economic considerations

Continuously, breeds and lines are set outside the primary food production chain because they can no longer compete in economic performance. As stated in previous chapters there is a variety of objectives for conserving them. However, in the real world, funds for conservation will only become available when economic arguments can be used and economic values can be assigned to the breed or lines to be conserved.

Drucker (2001) gave an overview of the different economic values that may apply. He stated that the total economic value (TEV) of farm animal genetic resources could be expressed as follows:

$$TEV = DUV + IUV + OV + NUV$$

Where:

- *DUV* is *direct use value* and refers to the benefit resulting from actual uses, like food, wool, sports, draught, fertiliser, fur, etc. It is related to the objective to maintain breeds in rural areas as a socio-economic activity,
- *IUV* is *indirect use value* and reflects the benefits derived from ecological or cultural functions. It is related to the objective to maintain agro ecosystems in rural areas,
- *OV* is the *option value* and is derived from the value given to safeguarding an asset for the option of using it in future times. It is a kind of insurance value against the occurrence of, for example, a new animal disease or a climate change. This value is related to all objectives to maintain flexibility: an insurance for changes in markets or environments, a safeguard against disasters and the opportunities for research,
- *NUV* is *non-use value*, which consists of:
  - *Bequest values (BV)* which measure the benefit accruing to any individual from the knowledge that others might benefit from a resource in future. This value might be related to present socio economic objectives which also will be important for later generations,

- *Existence values (XV)* which are simply derived from the satisfaction of knowing that a particular asset exists, for example museum values of breeds. It is related to the cultural historic objectives for conservation.

It is important to note that current economic decisions are largely based on the direct use values alone. For sustainable use and conservation of genetic resources, the other categories may be of equal or greater importance, and are indeed likely to be positive (Drucker, 2001). By focusing exclusively on direct use values, biodiversity and genetic resource diversity are consistently underestimated. Rationally speaking, policies and decisions about farm animal genetic resources should be made in a way that the 'utility' of these resources is maximised. However, it would be a great challenge to assess the monetary values of those elements encompassing the total value defined above.

Clearly, the most frequently considered economic values are related to the direct use value. These rates of return are those that benefit the farmer directly, rather than society. A breeding program with maximum production per animal as its main goal may reduce values of the animal genetic resources that belong to the interest of the public and the consumers. This can be caused by:

- the existence of a negative genetic relationship between production traits on one side and functional traits and animal welfare traits on the other side;
- the possibility that a high intensity of selection may result in accumulated inbreeding, reduced genetic diversity and accumulation of deteriorating genes expressed as decreased fitness;
- the negative effect intensive production methods may have on the environment.

All the causes and effects mentioned above imply increased risk of disturbances in the production systems and reduce utility expressed as a satisfying service to society. Thus, there should be a real interest from the public and the consumers that the breeding schemes are run sustainable, which the government should secure.

The major characteristic for the evaluation of a breeding program is the economic response, which is, in fact a measure of the efficiency of the breeding program. The economic value of efficiency is based on the improvement of the breed's production process in monetary terms. The economic return on investment of breeding programs is large. It is estimated that the benefit/cost ratio of breeding programs in farm animals varies between 5:1 and 50:1 (Barlow, 1983; Mitchell *et al.*, 1982). These figures hold only true when the long-term total effects aiming at food security are positive. The calculated benefits from the breeding program are only accurate and will only be realised if the sustainability requirements for all resources involved are met (Cunningham, 2003).

Using breed or breed properties as part of a trademark value for the product of this breed means that the return on investment is greater when based on a higher price of the product than on the improvement of the production process leading to lower costs. Practically, the efficiency of the production process is 'paid' by lower cost, but the value of the product may be regulated by the price mechanism through the demand of a niche market for special products. This complicates the calculations of return on investments for breeding programmes.

4. Indicators for sustainable management

Sustainable management of a breed may be described by the following factors:
- utilisation and conservation of genetic diversity;
- the existence of a sustainable breeding goal;
- implications of interactions between genotype and production systems are taken into account, i.e., the adaptation of the breed is facilitated;
- food security and food safety are maintained at required standards;
- no environmental impact.

4.1. Utilisation and conservation of genetic diversity

A breeding scheme is a set of procedures that organises available breeding technology into a decision tool to optimise selection. The contribution theory put forward by Woolliams and Thompson (1994) presents a fundament for balancing selection intensity and inbreeding. The studies by Meuwissen (1997) and Grundy *et al.* (1998) produced tools to maximise the rate of genetic progress while restricting the rate of inbreeding. The direct effect might be that the effective size of the population may be kept constant by the restriction put on the rate of inbreeding. The indirect effect of the method of restricting the inbreeding rate is that it maximises the probability of selecting breeding parents who contribute genes and gene combinations in the random (Mendelian) sample term of the individuals in the next generation, that have never been expressed in the previous generation of parents or their ancestors. This means that each individual sampling term possesses and expresses a set of gene combinations and genetic effects that are unique for that particular individual in the population. Thus, the new parents contribute in a maximum way to the re-establishment of the genetic variation of the individuals in the next generations. It has been shown that a breeding program improves genetic selection efficiency by 20 to 25 % at the same inbreeding rate when an appropriate restriction on inbreeding is used (Avendãno *et al.*, 2004). These findings may easily convince breeding organisations to use the approach immediately, due to the improvement of efficiency it ensures and the long-term genetic progress it enables.

Despite the use of these methods, founders might be lost in future generations. Therefore, it is recommended to store genetic material from founders in a cryobank. Besides, it has to be realised that the continuous genetic progress raises the genetic level of the actual population, which may be quite different from the founders. Therefore it is also recommended to store a sample of the genetic material of subsequent generations.

Considerable genetic diversity comes from between-breed variation. In order to maintain this diversity for future use, efficient conservation schemes on national and global levels must be established. From animal breeding theory, it is known that immigration of genes is very efficient in re-establishing diversity by arresting any accumulated inbreeding, and by increasing the effective population size at the same time. This procedure can be used as long as there exist alternative breeds, from which genes can be imported. However, it assumes that a breed society accepts the use of such breeding methods. This illustrates that *maintaining several breeding populations, at least at the global level, is necessary to ensure access to a diversity of breeds.*

Immigration of genes has been used in Norwegian Red Cattle for years. In order to use an alternative breed on a regular basis, the national breeds of Swedish Red, Finnish Ayrshire and Danish Red must be maintained as individual breeds. Such a cooperation policy is a long-term insurance for the relevant societies, consumers, farmers and breeding organisations.

For the last 2-3 decades, local Friesian populations were upgraded by using Holstein Friesian sires from North America, resulting in one global Holstein Friesian population. This upgrading process resulted in the loss of the insurance provided by the diversity contained in the various national Friesian breeds or lines. Since this insurance of having different local Holstein Friesian breeds is lost, a new strategy should be outlined, to secure the size of the effective population at local levels or even globally.

Strong selection in a breed may lead to the loss of the so-called private alleles (Foulley and Ollivier, 2006). The local breeds seem to have conserved more private alleles than the commercial breeds. This might have an effect on the option values of local breeds.

These examples show that there are no international policy rules to maintain the "between commercial breed diversity" as a resource for future use. Today, the maintenance of this diversity can only be saved and further utilised by agreements and cooperation between breeding organisations from several countries. Such agreements start with political regulations that have to be implemented at the national level. The Swedish government's proposal of obligatory maintenance of genetic diversity, as part of its animal production act, might be good start for such a national approach.

4.2. A sustainable breeding goal as selection objective

The realised weights for the traits and the number of different traits in the breeding goal to be improved by selection are very important in a long-term selection programme. In addition, the correlated effects of traits that are not included in the breeding goal may be adversely affected and this actually will accumulate over time. Balancing production and functional traits becomes increasingly important as time passes. The main reasons are:

- In general, the genetic correlation between production and functional traits is negative. Thus, selection only for production traits results in negative effects on the functional traits.
- Positive and negative effects of breeding programmes accumulate over generations. For example, small negative changes per year accumulate and may appear after a few decades to be detrimental for welfare fitness traits like fertility, disease resistance and viability.

However, when taken into account, functional traits in the selection programme can yield positive results, as illustrated by the genetic trend for mastitis resistance shown in Figure 9.2.

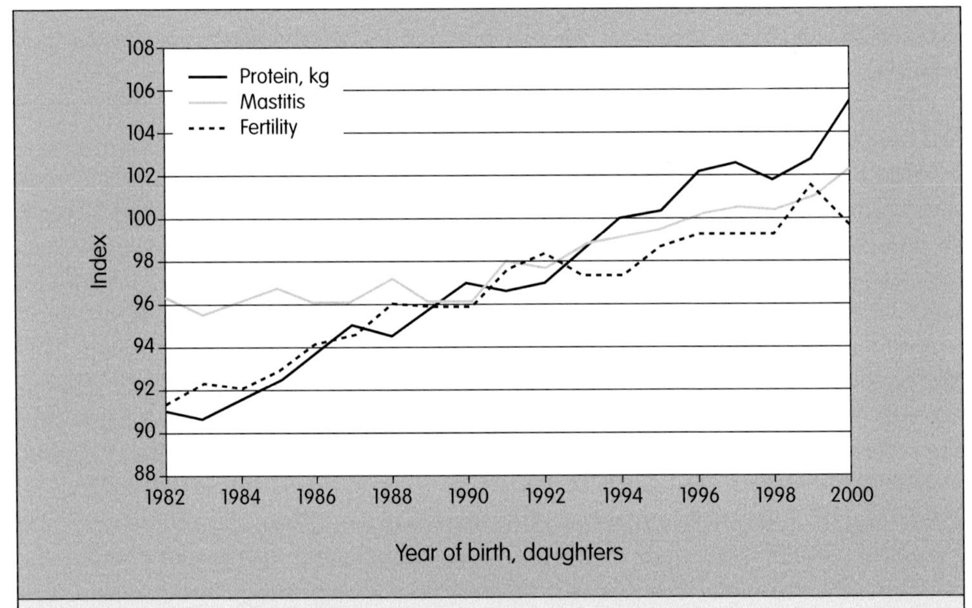

Figure 9.2. Genetic improvement of mastitis in Norwegian Red (Source: http://www.geno.no/genonett/presentasjonsdel/engelsk/default.asp?menyvalg_id=418 and go to: Norwegian Red characteristics).

The weight on mastitis in the total merit index was increased significantly from 1990 onwards. This has enhanced the positive trend considerably. The increase in genetic resistance to mastitis is really promising for farmers who wish to produce milk without using antibiotic treatment. This will indeed increase the safety of milk production and facilitates, for example, organic milk production. Figure 9.2 also shows the selection response of non-return rate and protein yield. An effect not shown in this figure is that other diseases than mastitis follow a similar genetic trend. In the search for a sustainable breeding goal, balancing the production traits with the fitness traits is an effective long-term selection strategy.

4.3. Implications of interactions between genotype and production systems are taken into account

When the testing of the breeding animals and the production of the offspring are performed in the same environment or in the same production system, the interaction between genotype and environment or production system can be ignored. However, when the offspring are exported, the new production environment may be quite different from the test environment of the exporting country. Besides, a lack of adaptation of the breeds to the environment in the target country might have a negative effect on fitness traits, and result in disappointing production figures. An international regulation of exchange of farm animal genetic resources should focus on the existence of possible interactions and their long-term social and economic consequences for the importing country. Ignoring the effect of this type of interaction might undermine the livelihood of farmers in the importing country. Such import often implies that the local livestock systems erode, and often the livelihood of entire groups of people is destroyed. As much as 70 % of the world's rural poor (approximately 2 billion people) keep livestock to meet the food demands of their families. For these people, livestock diversity thus contributes in many ways to human survival and well - being (Drucker, 2001).

4.4. Food security and safety

Woolliams (2005) discusses the fundamental importance of farm animal genetic resources for food security and safety. Livestock development works best when all strategies are co-ordinated and work in the same direction. For example, fertility in dairy cattle tends to decrease as milk yield increases. An established consequence of infertility is an increase in greenhouse gas emissions from the production system per litre of milk produced. The effectiveness of any management solution will be compromised when selection increases yield without taking into account the genetic merit for fertility. In this case, the overall utility of the system is not optimised (Woolliams, 2005). Genetics can play an important role in the dynamics of the populations resulting from genetic selection, and one should use genetic options, where they exist, as part of the solution to improve security and safety.

To meet the challenges to food security arising from increased global demand and the threats from global warming, livestock breeding must be included as a component of the solution.

4.5. Effects on the environment

Animal production can imply positive or negative effects on the environment. In the most extreme production circumstances, the value of the environment for society might be reduced.

Increasing the production volume may also increase waste output. The considerable volumes of waste produced by large-scale, high-density livestock operations can cause severe soil, water and air pollution (Cunningham, 2003). The emissions giving rise to most concern are nitrogen, phosphorous, various heavy metals and greenhouse gasses such as methane and nitrous oxide. If the recycling of manure and urine to agriculture is not firmly regulated, considerable environmental damage may arise. The strong focus on environmental issues in several countries may lead to regulations that minimise the output of wastes from livestock systems. Such regulations may require other genotypes than those favoured by the present breeding goals. This means that breeding programmes that maximise production volume per animal may lead to a reduction of environmental quality for society.

5. Monitoring role for the national managers

Monitoring farm animal genetic resources is an essential activity for sustainable utilisation and conservation, provided that policies are being addressed and a national manager or co-ordinator is appointed, i.e., within a coordination unit. This manager can act as a reference point in times of crisis within the livestock industry, e.g. the outbreak of an emerging disease. The manager measures regularly the degree to which targets have been achieved and identifies at an early state possible new emerging problems. This determines the context within which policies are reviewed and refined.

In the book *Sustainable Management of Animal Genetic Resources* (Woolliams *et al.*, 2005), monitoring functions were outlined. The breeding companies and a national centre of genetic resources might easily perform the monitoring activities for farm animal genetic resources. The latter should annually provide the monitoring results to the responsible national authority.

The key factors for monitoring the sustainable utilisation and management of national farm animal genetic resources include:

- The estimated value of the effective size of the breeding populations (breeds) linked to the actual rate of inbreeding.
- Estimated genetic trend for production traits and all important fitness traits.
- The trend of the number of breeding females of the endangered breeds.
- The priority of endangered breeds according to their category of threat.

6. Sustainable breeding schemes

In recent years, breeding organisations became aware of the societal interest in their work. With input from producers, farmers' organisations, NGOs and policy makers, the European Forum of Farm Animal Breeders developed a Code of Good Practise for Farm Animal Breeding and Reproduction (EFFAB, 2005). This code presents the principles of conduct for animal breeders, who are at the beginning of the food chain and thus directly responsible for providing quality genetics to farmers. These animal breeders operate on a highly competitive global playing field. Therefore, a sustainable solution balancing technical and economic realities, animal welfare, genetic diversity and public opinion is necessary to remain competitive in the future. The sustainability codes comprise: food safety and public health, product quality, genetic diversity, efficiency, environment and animal welfare and health. A checklist for sustainable breeding schemes is outlined in Box 9.3. The statement on genetic diversity is worded as: "Breeding programmes are designed to make optimal use of existing genetic variation between and within populations and to control inbreeding.

7. Future policies for animal genetic resources

As stated in chapter 1, the increasing trade of animal products and genetic material across countries, regions and continents jeopardises food security and food safety. The threat to food security is caused by the loss of local or regional breeds and to some extent to increased inbreeding within pure breeding populations.

In the past, most of the genetic material from farm animals was owned by the (national) organisations of farmers. During the past 15 years, great changes in the ownership of such material have occurred, especially within the poultry, but also within the pig sector. A limited number of multinational poultry breeding companies, who also own the breeding stock, are now providing the genetic material for the intensive egg and broiler production systems all over the world.

Box 9.3. Checklist for sustainable breeding schemes (Woolliams *et al.*, 2005).

- Are market and product well defined? This includes:
  - A definition of the production system(s) including the restrictions on inputs and wastes, straight-bred or crossbred.
  - Expectations for trends in political, economic and social attitudes.
  - An assessment of the need for marketing.
- Is the breeding goal well defined? This includes consideration of:
  - Both income and costs of production within system(s).
  - Animal health and welfare.
  - Documentation and review procedures.
  - Acceptance by consumers, producers and breeders.
- Is sensitivity to environmental factors addressed? This includes:
  - Fluctuations and trends in the market.
  - Backup to account for unexpected situations such as diseases and accidents e.g. cryoconserved germplasm, dispersed elite populations.
  - Food safety.
  - Genotype-environment interactions.
  - Consumer and producer acceptance of the measurement and reproduction techniques used.
- Are sufficient economic, technical (including R&D) and human resources available?
- Can livestock resources and selection strategies secure a sufficiently large effective population size to keep ΔF under 1% per generation?
- Is recording sufficient?
  - To obtain response in all components of the breeding goal.
  - To detect undesirable changes in animal health and welfare.
- Are expected effects of selection predicted? This includes:
  - The genetic trends for all traits in the breeding goal, or potential important traits not in the breeding goal.
  - The predicted impact of changes in recording.
- Is genetic progress monitored and evaluated?
- Are time horizon and milestones defined? This includes:
  - Predicted and realised genetic progress.
  - Markets and breeding goal review.
  - Costs and benefits of the breeding scheme to the breeders, the producers and consumers.
- Is the profitability of the breeding scheme evaluated?

Such trends lead to an increasing threat to our food supply. This might force the global society to develop a 'regulation system' for the exchange of genetic material. Such a 'regulation system' should be based on such principles as:

- The exchange or trade of genetic material should be stimulated when it gives benefit for the sellers and the buyers and when it contributes to sustainable development of an animal population in the new production environment.
- The 'regulation system' should put responsibility on both the sellers and the buyers to secure that the genetic material transferred is adapted to the new production environment. When the genetic material has not been exposed to the new environment earlier, the seller should develop a test program to avoid failures that might give large damage for the local production system and the local breeds in the importing country.
- The 'regulation system' should not require large transaction costs to manage the transfer of the genetic material.
- The 'regulation system' may be based on a standard transfer agreement, which may be linked to the national jurisdictions of the receiving country, or may be a multinational agreement signed by the parties involved.
- The 'regulation system' may be a formal treaty similar to the FAO's Treaty on Plant Genetic Resources used in agriculture. However, a treaty for animal genetic resources should not be a copy of Treaty on Plant Genetic Resources, because of the large differences in ownership, technical differences and the different requirements for conservation *in situ* and *ex situ*.

In a transfer agreement it should be clarified whether the transferred material is used for food and agricultural production only, or for breeding purposes. The latter implies further use for inventions and may raise problems related to intellectual property rights through the patenting system.

The scope of use of the transferred material must be clearly determined in the material transfer agreement, which may require negotiations between the parties to agree on benefit sharing of the added values obtained from the transferred material after an invention.

8. Sustainability issues for the different stakeholders

There is a real need for a better understanding of the sustainability issues, where the stakeholders have different perspectives. In the short term, the different stakeholders in animal breeding can have very diverging objectives: (1) animal breeders and producers focus mainly on production and the factors that may reduce the costs of primary production; (2) manufacturers and retail groups focus on product quality, quantity, the profitability of manufacturing, and adding value to the primary product; (3) consumers and the general public are primarily concerned with issues of product quality, safety, prices, cultural and political issues, including animal welfare, environmental impact and pollution. In the long-term, the preferences of society dictate the broad framework supporting the production of animal products and, as the urbanisation of society continues, these preferences will become increasingly aligned with those of the consumers. Therefore, despite the differences in short-term objectives, stakeholders do share many long-term objectives for securing the sustainable production of animal products (Woolliams *et al.*, 2005).

The complexity of these issues requires a thorough analysis and serious discussions and negotiations between stakeholders, including the national, regional and global political levels.

References

Avendãno, S., J.A. Woolliams and B. Villanueva, 2004. Mendelian sampling terms as a selective advantage in optimum breeding schemes with restrictions on the rate of inbreeding. Genetical Research 83: 55-64.

Barlow, R., 1983. Benefit-cost analyses of genetic improvement program for sheep, beef cattle and pigs in Ireland. Ph.D. Thesis, University of Dublin. Ref. by E.P. Cunningham, Present and future perspective in animal breeding research. XV. International Congress of Genetics, New Delhi, India, 12-21 December 1983.

Cunningham, E.P. (ed.), 2003. After BSE – A future for the European livestock sector. EAAP publication No. 108. Wageningen Academic Publishers, Wageningen, The Netherlands.

Drucker, A.G, 2001. The Economic Valuation of AnGR: Importance, Application and Practice. In: Proceedings of the workshop held in Mbabane, Swaziland, 7-11 May 2001.

EFFAB, 2005. Code-EFABAR. Code of Good Practise for Farm Animal Breeding and Reproduction (FOOD-CT-2003-506506). www.code-efabar.org

FAO, 2004. The Legal Framework for the Management of Animal Genetic Resources, Background Study Paper NO. 24. ftp://ftp.fao.org/ag/cgfra/BSP/bsp24e.pdf

Fimland, E., 2006. Institutional issues and frameworks in *ex situ* conservation of Farm Animal Genetic Resources (FAnGR). J. Gibson, S. Gamage, O. Hanotte, L. Iñiguez, J.C. Maillard, B. Rischkowsky, D. Semambo and J. Toll (eds.), 2006. Options and Strategies for the Conservation of Farm Animal Genetic Resources: Report of an International Workshop and Presented Papers (7-10 November 2005, Montpellier, France) [CD-ROM]. CGIAR System-wide Genetic Resources Programme (SGRP)/Biodiversity International, Rome, Italy.

Foulley, J.-L. and L. Ollivier, 2006. Allele Diversity. Proceedings 8th World Congress on Genetics Applied to Livestock Production, CD-ROM Communication No. 33-09.

Grundy, B., B. Villanueva and J.A. Woolliams, 1998. Dynamic selection procedures for constrained inbreeding and their consequences for pedigree development. Genetical Research 72: 159-168.

Mitchell, G.C., M. Smith, P.J. Makower and W.N. Bird, 1982. An economic appraisal of pig improvement in Great Britain. Genetic and production aspect. Animal Production 35: 215-224.

Meuwissen, T.H.E., 1997. Maximising the response of selection with a predefined rate of inbreeding. Journal of Animal Science 75: 934-940.

NCM, 2003. Access and Rights to Genetic Resources. A Nordic Approach. Nord 2003:16. Nordic Council of Ministers, Store Strandstrede 18, DK-Copenhagen, Denmark.

Woolliams, J.A., 2005. Sustainable Livestock Breeding – Food Security and Safety. Nordic GENEresources, http://www.nordgen.org/publikasjoner/nordiskegenressurser.htm

Woolliams, J.A. and R. Thompson, 1994. Proceedings of the 5th World Congress on Genetics applied to Livestock Production 19: 127-134.

Woolliams, J.A., P. Berg, A. Mäki-Tanila, T.H.E. Meuwissen and E. Fimland, 2005. Sustainable Management of Animal Genetic Resources. Nordic Gene Bank Farm Animals.

Glossary

Adaptation is a particular change or a set of changes in the abilities of an individual, or a population, that increases its fitness in its environment. See 'Fitness'.

Additivity is the assumption that each allele influencing a trait does so independently of the other allele present at that locus and all other alleles at all other loci, e.g. if alleles Q and q are worth 1 and -1 respectively then additivity assumes QQ is worth 2, Qq is worth 0, and qq is worth –2.

Admixture is a population with a mix of two ancestral groups, c.f. a synthetic in livestock breeding. Populations with admixture will display comparatively large DNA (marker) diversity with comparatively greater linkage disequilibrium. The observable extent of this disequilibrium will decrease over generations achieving enough recombinants for fine mapping. See 'Crossbreeding'.

Admixture mapping is a method for fine mapping and localising QTL (e.g. a disease causing allele) when the trait (e.g. disease incidence) differs across populations, and where an admixture of these populations exists. The approach assumes that near the disease-causing allele there will be enhanced ancestry from the population that has greater risk of getting the disease, and ancestral origins over short distances can be identified from the additional linkage disequilibrium in the admixture.

Allele is a version of the sequence of DNA nucleotides at a locus. Not all individuals carry exactly the same sequence of DNA nucleotides at a locus. This allelic variation is the source of genetic variation e.g. the phenomenon of variation in double muscling in cattle is due to there being two versions of DNA nucleotides at the locus that codes for a protein called myostatin.

Backcross is a cross produced by mating a cross formed from matings of two lines or breeds back to an individual from one of the founding lines or breeds.

Bayesian is an approach to statistical inference that assumes parameter values are random variables, with prior distributions describing our strength of belief in possible values before the collection of data. In contrast, a frequentist approach assumes parameters are unknown constants. These different starting points result in different philosophics of inference after the collection of data, with Bayesian inference based upon the posterior distributions for parameters, and frequentist inference based upon confidence intervals and hypothesis testing calibrated by hypothetical repetition of the data collection.

BLUP is an acronym for Best Linear Unbiased Prediction. It is a standard statistical method for estimating breeding values in populations in an optimal way. BLUP accounts for genetic relationships and adjusts for systematic fixed effects simultaneously.

Bootstrapping is an analysis in which multiple datasets are formed by random sampling with replacement from an actual data set, in order to estimate a degree of confidence in an estimated parameter. This process is used widely in QTL mapping and in phylogeny reconstruction.

Bottleneck is a period when the number of parents used to reproduce the breed was particularly small. In such a period the genetic drift is high due to a marked reduction in the effective size of the population.

Breed is an interbreeding group of animals within a species with some identifiable common appearance, performance, ancestry or selection history; see box 3.1 for more details.

Breeding organisation is a term to represent all organisations involved in livestock improvement: breeding companies, breed societies and breeders collaborating in group-breeding schemes.

Breeding objectives or goals represent the direction of change desired within the population. Very often these objectives are limited by what records are available for evaluation, and an organisation will define its objectives by what it *can* do rather than what it would do if records existed, although this is not best practice. New opportunities will expand the achievement of objectives in practice.

Breeding programme or a breeding scheme is a programme aiming at defined breeding objectives for the production of a next generation of animals. It is the combination of recording selected traits, the estimation of breeding values, the selection of potential parents and a mating programme for the selected parents including appropriate (artificial) reproduction methods.

Breeding value is the mean genetic value of an individual as a parent, for one trait or a combination of traits. It is estimated as twice the average superiority of the individual's progeny relative to all other progeny under conditions of random mating. An individual's breeding value defines its additive genetic value.

Candidate gene is a gene chosen among genes known to affect a studied trait or among genes in a QTL region found as an important source of variation in a genome screen. See 'Positional candidate'.

Centi-Morgan is a linkage map distance of 0.01 Morgans, corresponding roughly to 1 per cent recombination; abbreviated cM. See 'Morgan'.

Chromosome is a discrete block of DNA and is one of the basic structures of the genome. All nuclear DNA is organised into chromosomes with the number varying between animal species. Genes on a chromosome are linked and tend to be inherited together.

Clone (animal) is an individual that is genetically identical to another or a group of individuals that are genetically identical to each other.

Co-ancestry is the relationship by a common ancestor of both one's father and one's mother; synonym for kinship coefficient.

Co-dominance is a situation in which a heterozygote shows the phenotypic effects of both alleles equally. See 'Additivity'.

Comparative genomics is a joint analysis of the genome between two or more species, making use of known similarities between the structures of their genomes.

Conservation potential is the marginal diversity multiplied by the extinction probability. It reflects the benefit in terms of conserved diversity of making a breed completely safe.

Core set is the smallest set of breeds or lines of a species that still encompasses the genetic diversity in that species.

Crossbreeding means matings between animals of different breeds or lines.

Cryoconservation or **cryopreservation** is the maintenance of germplasm in the form of tissues, semen, oocytes, or embryos in long-term storage at ultra-low temperatures, typically between -150 and -196 Celsius in liquid nitrogen, for the purpose of subsequent use to produce viable live animals.

Cytoplasmatic inheritance is the transmission of hereditary traits through self-replicating factors in the cytoplasm, for example: mitochondria and chloroplasts.

Diploids carry two sets of chromosomes. With the exception of sex chromosomes, diploids carry 2 copies of each locus and 2 copies of like-structured chromosomes.

DNA is Deoxyribonucleic Acid, which is a macromolecule in the form of a double-stranded helix that carries the genetic information in all cells in higher organisms.

Domestication is the process in which animal populations adapt to mankind and its environment. It may be also considered as a form of mutualism involving a parallel evolution in culture and genome. Animals such as dogs, pigs, cows, and sheep were domesticated from their wild relatives by humans thousands of years ago.

Dominance is when the alleles of a locus are non-additive. When a locus shows dominance, the genotypic value of the heterozygote on a trait is not the average of the two homozygotes. Overdominance occurs when the heterozygote has a genotypic value more extreme than either parent. See 'Recessive allele'.

EBV is an acronym for Estimated Breeding Value.

Ecosystem is the complex of a living community of species and its environment, functioning as an ecological unit in nature.

Effective population size (Ne) for a population is the number of diploid, single-sex individuals that when randomly selected and randomly mated (including selfing) that would be expected to have the same rate of inbreeding as the population itself. See chapter 3.6.

Epistasis is when loci are non-additive. The genotypic value of a locus on a trait depends upon the genotypes at other loci or a situation in which the differential phenotypic expression of a genotype depends on the genotype at another locus.

Evolutionary tree is a diagram of the inferred ancestry and descent among a group of species or populations. Within species, a tree assumes that sub-populations, once diverged, never mix.

***Ex situ* cryoconservation** see 'Cryoconservation'.

***Ex situ in vivo* conservation** or ***ex situ* live conservation** is defined as conservation by maintaining a live population either under abnormal farm conditions, or outside of the area in which it evolved or is now normally found, or both: e.g. when a few animals of a breed kept in zoos or farm parks for cultural or historic reasons. The costs of this type of conservation are low, but further adaptation of the population to the native environment is impossible.

Extinction probability is the probability that a breed will go extinct within a defined time horizon (e.g. within the next 25 years).

Factorial mating is a mating scheme where each male is mated to more than one female, *and* each female is mated to more than one male. Such mating schemes can either be partial or complete; the latter being when each parent is mated to all parents of the opposite sex. In some species, this is made more tractable by means of *in vitro* embryo production. Such a mating scheme substantially reduces the rate of inbreeding in genetic improvement.

Fitness is an important genetic concept and is the trait defined by the relative number of offspring left by an individual compared to its competitors. Whilst artificial selection in improvement schemes influences this process, fitness is viewed as a composite of all traits involving health and well-being influencing the ability of an individual to survive and to produce viable offspring.

Founder is a term for an individual in the base generation of (typically) a conservation scheme. It too has a pedigree, perhaps unknown, that was subject to genetic drift, migration, selection, and mutation, and will have offspring and grand-offspring and later descendants that will form the next generations, and the management of these will then be more critical to the future gene pool than the founders. Founders should not be idolised. Genetic drift and hence change in allele frequencies can occur whilst still sampling founders absolutely equally.

Gamete contains one haploid set of chromosomes passed from a parent to an offspring. So in diploid species, the offspring receives 2 gametes, one from the sire and one from the dam.

Gene is the hereditary unit, a region of DNA on a chromosome containing genetic information that is transcribed into RNA that is translated into a polypeptide chain with a physiological function. A gene can mutate to various forms called alleles. See 'Allele'.

Gene bank is an institution or centre that participates in the management of genetic resources, in particular by maintaining *ex situ* or *in situ* collections; the term can also refer to a collection of genetic resources rather than the institution holding it.

Generation interval is the period of time taken to renew the population of parents. The definition for male and female parents is the average age of the parent when its replacement is born. The generation interval for the population is then the average of these two values since males and females each contribute half the genes to renewing the population.

Genetic distance is a measure of the genetic similarity between any pair of populations. Such distance may be based on phenotypic traits, marker allele frequencies or DNA sequences.

Genetic diversity is the set of differences between species, breeds within species, and individuals within breeds expressed as a consequence of differences in their DNA.

Genetic drift is the change in the frequency of alleles in a population resulting from sampling variation in drawing gametes from the gene pool to make zygotes and from chance variation in the survival and or reproductive success of individuals.

Genetic erosion is a permanent reduction in the number, evenness and distinctness of alleles, or combinations of alleles, of actual or potential agricultural importance in a defined geographical area.

Genetic improvement is a change in the genetic capability of a population directed towards its breeding objectives.

Genetic load is the reduction in the mean fitness of a population due to the presence of deleterious alleles.

Genetic marker is a specific and identifiable sequence of DNA (see also Box 3.3)

Genetic resources are the carriers of the genetic variation.

Genetic variation is a statistical measure of the extent of differences among individuals in a population that is due to differences in genotype.

Genome is a collective term for all DNA in the cell nucleus i.e. the set of chromosomes.

Genome bank. See 'Genebank'.

Genotype is the pair of alleles of an organism carried at a locus. The term is sometimes used to mean the set of genotypes at all loci being considered.

Genotype-environment interaction occurs when the difference in performance between two genotypes depends on the environment in which the performance is measured. This may be a change in magnitude of the difference or a change in the rank of the genotype.

Germplasm comprises the tissue, semen, oocytes, embryos, or juvenile or mature animals useful in breeding, research and conservation efforts.

Germplasm bank. See 'Genebank.

Haploids carry one set of chromosomes.

Haplotype is a combination of alleles over (closely) linked genes or markers carried on a single chromosome. Haplotypes therefore tend to be inherited as a unit, but change over generations by recombination.

Hardy-Weinberg equilibrium occurs at a locus after one generation of random mating, and shows characteristic frequencies for the homozygotes and heterozygotes depending on the overall frequencies of the alleles. These expectations can be used to test for the presence of non-random mating. Preferences for mating relatives increases the frequencies of homozygotes above the number expected, whilst avoidance of relatives in mating increases the frequency of heterozygotes.

Heritability is the fraction of phenotypic variance that is attributable to genetics. The genetic variance used is most commonly the additive genetic variance i.e. the variance of breeding values.

Heterosis or hybrid vigour is the extent to which the performance of a crossbred in one or more traits is better than the average performance of the two parental populations. This is an expression of epistatis or dominance. See 'Epistasis' and 'Dominance'.

Heterozygote is an individual carrying two distinct alleles at a locus, e.g. Qq.

Hitch-hiking is the change in the frequency of an allele due to selection on a closely linked locus with a positive allele.

Homologous describes when two segments of DNA fulfil the same purpose in the genome; therefore diploid individuals, such as mammals, have chromosomes in 'homologous' pairs, one version inherited from the sire and one from the dam, similarly 'homologous' alleles.

Homozygote is an individual carrying two copies of the same allele at a locus, e.g. qq or QQ.

IBD is Identity-by-Descent. Each offspring receives a copy of one of the two alleles carried by each of its parents. Two alleles are identical by descent if, when traced back over generations, are copies of the same allele carried by an ancestor.

Inbreeding is the formation and accumulation of loci that are IBD, arising from the mating of parents with a common ancestor, which is inevitable over long periods of time. It is measured by a probability that a locus is IBD.

***In situ* conservation** is defined as conservation of a livestock population through continued use by livestock keepers in the agro-ecosystem in which the livestock evolved or are now normally found (includes breeding programmes).

Introgression is transfer of an allele or set of alleles from one breed to another. This is achieved by the crossing of a number of parents from the donor breed to the recipient breed, followed by systematic backcrossing to the recipient breed, with parents chosen to be carriers of the desired alleles. Markers can be used to detect these carriers.

Kinship coefficient is a probability of IBD when sampling (with replacement) an allele from the same locus in two individuals.

Linebreeding is the mating of selected individuals from successive generations to produce animals with a high relationship to one or more selected ancestors. It is a form of inbreeding. In livestock breeding it is used to develop potentially advantageous production traits within a group of individuals maintained in reproductive isolation. The developed lines are mainly used in crossbreeding programmes.

Linkage is the phenomenon by which alleles at loci that are close together on a chromosome and which have been inherited together from one parent of an individual tend to be passed on together to an individual's offspring. The closer the loci are on a chromosome the stronger is this phenomenon. When the loci are on different chromosomes then this tendency is completely absent.

Linkage disequilibrium is a non-random association of alleles in haplotypes. Over time recombination events between loci will remove this association, more quickly the further away the loci are from each other.

Locus is a position in the genome i.e. a position on a chromosome. The plural is loci.

Marginal diversity is the change of conserved diversity at the end of the considered time horizon, if the extinction probability would be changed by one unit by a conservation effort.

Marker-Assisted-Selection (MAS) is selection for a trait of interest where selection criteria include the genotype(s) of linked testable genetic marker(s).

Mating systems are the rules that describe how selected breeds or lines or individuals will be paired at mating.

Matrilinear diversity is diversity found when tracing descent through the female line See 'mtDNA'.

Meiosis is the process carried out in the germ cells by which gametes are formed. In diploids this involves the creation of haploid cells (sperm, oocytes) from the diploid progenitor cells.

Mendelian sampling is the random sampling of parental genes caused by segregation and independent assortment of genes during germ cell formation, and by random selection of gametes in the formation of the embryo.

Morgan is a map distance on a chromosome, defined by the expected number of crossovers occurring during meiosis between the loci.

mtDNA is mitochondrial DNA. The mitochondria in the cells of individuals descent from the mother. It is a form of maternal extra-nuclear (cytoplasmatic) inheritance of a trait.

Mutation is an event that creates a change in the DNA sequence on a chromosome of an individual so that the sequence is not the same as that inherited from either sire or dam. In genetics this has most impact when the mutation occurs in germ cells so that it is passed to offspring. Mutational events are caused by irregularities in cellular processes, and when the mutation alters the function of the sequence in which it occurs it may introduce new genetic variation into the population.

MVO core set is a core set in which the quantitative genetic variance is maximised in a hypothetical population randomly bred from all the entities in the set.

MVT core set is a core set that maximises the total variance within and between populations.

Natural selection is the process of evolutionary adaptation in which those better suited to survive and reproduce in a particular environment give rise to a disproportionate share of the offspring. Where this fitness has a genetic basis, and where there is additive genetic variation for fitness, the overall ability of the population to survive and to reproduce in that environment will increase. See 'Fitness'.

Neutral loci are loci that are not evolving directly in response to selection, the dynamics of which are controlled mainly by genetic drift and migration. These loci can, however, be influenced by selection on nearby (linked) loci.

Non-additivity is when the additivity assumption fails and includes both dominance and epistatis.

Nucleus breeding scheme is a breeding scheme where a high level of recording is made upon a sub-population that is small proportion of the total population, so that more accurate and intense selection may be applied. The genetic improvement realised is disseminated into commercial populations.

Optimal contribution selection is a selection method that uses the average kinship of the selected parents to manage genetic variation. This can be implemented in various forms, such as maximising gain with a fixed rate of inbreeding, or minimising the loss of genetic variation.

Overdominance see 'Dominance'.

Pedigree is the set of known parent-offspring relationships in a population, often displayed as a family tree diagram. This can be used to derive the relationships and kinship coefficients between all individuals in the population. See 'Relationship' and 'Kinship coefficient'.

Phenotype is the observed value of a trait. It is a consequence of all the genetic and environmental influences and interactions affecting the trait, including errors in measurements.

Phylogenetic tree. See 'Evolutionary tree'.

Phylogeny is the evolutionary history of a population.

Pleiotropy (pleiotropic) is when a locus has an effect on more than one trait, for example the double muscling locus has effects on muscling score and calving interval, some of the loci affecting milk yield affect mastitis or fertility.

Polymorphism (polymorphic) is when the two alleles carried by an individual at a locus (one inherited from the dam, one from the sire) are different. See 'Heterozygote'.

Positional candidate is a locus that lies within a region of DNA that is known to harbour a QTL for a trait, and so may prove to be the locus with the causal mutation. A functional positional candidate is where (part of) the function of the positional candidate is known, perhaps from mapping projects in other species, and is considered relevant to the trait.

Preservation is that aspect of conservation by which a sample of an animal genetic resource population is designated for an isolated process of maintenance *in situ* or *ex situ*.

Private allele is an allele found only in one sub-population or breed.

Progeny testing is the evaluation of a genotype of a parent by a study of its progeny under controlled conditions.

Protected Designation of Origin (PDO) represents institutional brands that help to assure consumers that the product comes from a particular geographic area and was obtained using well defined quality standards. This origin labeling is usually managed by consortia which have the rights for management and promotion.

QTL is a Quantitative Trait Locus, a discrete, small segment of DNA that has a large effect upon a trait. This is in contrast to the traditional assumptions made in much genetic theory where it is considered that there are many, many loci influencing a trait each with a small effect upon it.

Random mating is a mating system in which animals are assigned at random as breeding pairs without regard to genetic relationships or performance.

Recessive allele is an allele that is only has an affect on the phenotype when it is homozygous. Therefore if allele q is recessive, qq yields a different phenotype from Qq and QQ, which have the same phenotype. Q is said to be the dominant allele. It is an example of non-additive gene action. See 'Dominance'.

Recombination occurs between a haplotype inherited from an individual's sire and the corresponding haplotype inherited from its dam. The individual passes a recombined haplotype to an offspring when a crossover occurs, i.e. the initial sequence of alleles is inherited from one parent followed by a sequence of alleles inherited from the other. The probability of a crossover depends on the length of the haplotype. Recombination erodes linkage disequilibrium.

Relationship or **relationship coefficient** has a technical meaning as the covariance between the breeding values of two individuals, scaled by the additive genetic variance. This can be shown to be equal to twice the kinship coefficient between the individuals.

RNA is Ribonucleic Acid, a nucleic single-stranded acid. See also 'Gene' and 'DNA'.

Selection footprint See 'Signature of selection'.

Selection index is a combination of measurements of several sources into an estimate of genetic value. It may include more than one measurement of the same trait and measurements of a trait on relatives and may combine more than one trait in an overall genetic value.

Selection intensity is the proportion of animals selected relative to the total number available for selection. The smaller the proportion selected, the higher the selection intensity.

Selective sweep. See 'Hitch-hiking'.

Sexed embryos or **sexed semen** are respectively embryos (sperm) separated according to sex by testing for the presence of X or Y chromosomes. This is achieved by a variety of means.

Signature of selection is the pattern of reduced diversity adjacent to a gene that has been strongly selected for or against within a population.

Sire reference scheme is where a number of sires have progeny in more than one herd or flock to facilitate breeding value estimations. This may be achieved by natural mating or by artificial insemination.

SNP is a Single Nucleotide Polymorphism caused by a mutation at a single nucleotide (in contrast to a deletion or other mutational event).

Speciation is the process of forming new species by the splitting of an old species into two or more new species incapable of exchanging genes with one another.

Species is a group of organisms that can exchange genes among themselves but are reproductively isolated from other such groups.

Sustainability is the ability to provide for the needs of the world's current populations without damaging the ability of future generations to provide for them.

Upgrading is a crossbreeding system in which females of local breeds and their female offspring are systematically mated to sires of an exotic breed, so that over time the population will have a genome that is almost completely derived from the exotic.

Index

O

P

Q

R

S

U

Printed in the United States
by Baker & Taylor Publisher Services